Praise for *The*

D0361749

"Lynne McTaggert is the Malcolm Gladwell of new science, whose superbly written book, a blend of frontier discovery told in compelling stories, offers hope but also an action plan in these troubled times. Read this book and change your life and the lives of everyone around you."

—Jack Canfield, author of *The Success Principle*

"There is something on every page of this book to make you jump for joy. It's like things you've secretly suspected but figured must not be true, since no one else seems to think they are, are suddenly proven to be true! You don't even realize how thirsty your soul has been for another way of looking at the world until it's given to you and you can't stop drinking it in. Rarely does a book so tell it like it is."

—Marianne Williamson, author of *A Return to Love*

"We all want a better world, a better life, but many people don't have the courage to take that first step. *The Bond* is so full of wisdom, truth, and science that it whispers to your heart and screams at your brain: Reclaim your life and create a better way to live for your sake, and your children's, and your children's children's."

—Marci Shimoff, author of *Happy for No Reason*

"Lynne McTaggart has elegantly and boldly caused us to look at the world we are living in through a new set of lenses. Her truth in this book shook me to my core, inspired me in my soul and called me to action like no other book has ever done. Lynne is brilliant, and this book is a work of genius that can guide us back to ourselves and to each other."

—Lisa Nichols, author of *No Matter What!*

"*The Bond* brilliantly reveals the essence of who we are backed by rigorous new science and expressed with clarity and power. A visionary and an extraordinary scholar of the heart, she communicates a life-altering message that has the power to transform the way we see ourselves and therefore the way we see the world. Read this book. It will open your heart and change your life."

—Lynne Twist, author of *The Soul of Money*

"In *The Bond*, the brilliant and wise Lynne McTaggart gives us a life-changing blueprint for a new way to live. This astonishing book, based on discoveries from the new sciences, sings to our hearts and speaks to our minds. The vitally important message that this mind-blowing book delivers has the power to heal our relationships, our neighborhoods, and our world, inspiring us into a whole new future at a time we need it most."

—Debbie Ford, *New York Times* bestselling author
of *The Dark Side of the Light Chasers*

"A bold paradigm-buster of a book full of not only scientific revelation but also hope and inspiration, offering a road map on how to live in harmony with each other and with our truest nature, as givers and sharers."

—Janet Atwood, coauthor of the bestselling *The Passion Test*

"*The Bond* takes us inside a luminous reality of connectivity at every level of existence, where we witness the core of the evolutionary impulse. In the heart of this design we are able to recover pathways to deep empathy, social harmony, and wholeness. You will bask in the afterglow of Lynne McTaggart's brilliantly interpreted science and in the visionary depth, moral coherence, and healing power of *The Bond*."

—James O'Dea, social healing activist, former president of
Institute of Noetic Sciences, and author of *Creative Stress:
A Path for Evolving Souls Living Through Personal and
Planetary Upheaval*

"Capably weaves [her findings] into a seemingly certain scientific case for holistic living within a cosmic 'Bond.'"

—*Kirkus Reviews*

"Her discussion of the Bond encompasses quantum physics, genetics, evolutionary biology, neuroscience, astronomy, sociology, psychology, and a host of other disciplines. . . . [A]mple and fascinating examples of 'holistic,' altruistic, and uplifting behaviors."

—*Publishers Weekly*

"McTaggert has done it again! *The Bond* is her tour de force. Everyone loves to feel love, everyone craves deep connection, but we don't always know how to live it. This simple but powerful message from the frontiers of modern science is a milestone that will appeal to the skeptic as well as the zealot. It shows us how to recover wholeness and heal our world."

—Arjuna Ardagh, author of *The Translucent Revolution*

"In *The Bond*, Lynne McTaggart smashes the myth that connection between people is a touchy-feely business. Using compelling stories and solidly researched data, she demonstrates in no uncertain terms that we need each other, and that our ability to thrive and perhaps even to survive depend on making this reality a fundamental organizing principle for our lives. *The Bond* is fascinating, inspiring, and provocative, and McTaggart's eloquence and reader-friendly style make this journey not only informative but engrossing."

—Ocean Robbins, founder of YES! and author of
Choices for Our Future

"Lynne McTaggart has once again hit a home run for the human potential movement. *The Bond* is a compelling compilation of fascinating and enthralling discoveries that awaken us to the interconnectivity of all of life, woven all together in a way that inspires each of us to happily take up the task at hand: the midwifing of deeper levels of relatedness and care back into our lives and back into our world."

—Katherine Woodward Thomas, cofounder/coleader
of the Feminine Power Global Community

"Lynne has once again cared, dared, and shared a timely and transcendent message for the sake of all of us who are mysteriously bonded across the world."

—Dr. John Demartini, human behavioral specialist,
educator, and bestselling author of *Inspired Destiny: Living
a Fulfilling and Purposeful Life*

Also by Lynne McTaggart

The Intention Experiment:
Using Your Thoughts to Change Your Life and the World

The Field:
The Quest for the Secret Force of the Universe

What Doctors Don't Tell You:
The Truth About the Dangers of Modern Medicine (updated 2005)

Kathleen Kennedy:
Her Life and Times

The Baby Brokers:
The Marketing of White Babies in America

THE
BOND

HOW TO FIX YOUR
FALLING-DOWN WORLD

LYNNE McTAGGART

FREE PRESS

New York   London   Toronto   Sydney   New Delhi

*f*P

Free Press
A Division of Simon & Schuster, Inc.
1230 Avenue of the Americas
New York, NY 10020

First Free Press trade paperback edition June 2012

FREE PRESS and colophon are trademarks of Simon & Schuster, Inc.

For information about special discounts for bulk purchases,
please contact Simon & Schuster Special Sales at 1-866-506-1949
or business@simonandschuster.com.

The Simon & Schuster Speakers Bureau can bring authors to your live event.
For more information or to book an event, contact the Simon & Schuster Speakers Bureau
at 1-866-248-3049 or visit our website at www.simonspeakers.com.

Manufactured in the United States of America

1   3   5   7   9   10   8   6   4   2

Library of Congress Control No.: 2010049795

ISBN 978-1-4391-5794-7
ISBN 978-1-4391-5795-4 (pbk)
ISBN 978-1-4516-1052-9 (ebook)

For Bryan

Everything was being destroyed, . . . but it seemed to me that one new thing was made. A kind of . . . responsibility. Man for man. . . .[T]o bring that on to the earth again like some kind of a monument and everyone would feel it standing there, behind him, and it would make a difference to him.

Arthur Miller, *All My Sons*

THE
BOND

HOW TO FIX YOUR
FALLING-DOWN WORLD

LYNNE McTAGGART

FREE PRESS

New York   London   Toronto   Sydney   New Delhi

Free Press
A Division of Simon & Schuster, Inc.
1230 Avenue of the Americas
New York, NY 10020

First Free Press trade paperback edition June 2012

FREE PRESS and colophon are trademarks of Simon & Schuster, Inc.

For information about special discounts for bulk purchases,
please contact Simon & Schuster Special Sales at 1-866-506-1949
or business@simonandschuster.com.

The Simon & Schuster Speakers Bureau can bring authors to your live event.
For more information or to book an event, contact the Simon & Schuster Speakers Bureau
at 1-866-248-3049 or visit our website at www.simonspeakers.com.

Manufactured in the United States of America

1   3   5   7   9   10   8   6   4   2

Library of Congress Control No.: 2010049795

ISBN 978-1-4391-5794-7
ISBN 978-1-4391-5795-4 (pbk)
ISBN 978-1-4516-1052-9 (ebook)

For Bryan

Everything was being destroyed, . . . but it seemed to me that one new thing was made. A kind of . . . responsibility. Man for man. . . .[T]o bring that on to the earth again like some kind of a monument and everyone would feel it standing there, behind him, and it would make a difference to him.

Arthur Miller, *All My Sons*

CONTENTS

CONTENTS

PART IV:

TOOLS FOR A NEW WORLD

PROLOGUE

One chilly Saturday morning, I was standing in a drafty auditorium watching one of my daughters in the midst of a dress rehearsal for her drama class's annual production. A talented actress, she had been chosen for the lead part during the auditions, but a few weeks before the dress rehearsal had been shunted to a more minor role. I had never been able to discover the reason for the change—and my daughter refused to talk about it—until one of her friends let slip that, when a new director took over, another thirteen-year-old girl had lied about her acting experience in order to persuade him that she should be given the part that had been assigned to my daughter—her best friend.[1]

When I tried to raise this tactfully with her mother, another spectator that day, she cut me off and shrugged. "Well, that's life," she replied airily, "isn't it?"

I was taken aback, but I had to admit she had a point. Certainly that's the life we grown-ups have designed for ourselves. Competition makes up the very warp and woof of the societies of most modern developed countries. It is the engine of our economy, and it is assumed to be the basis of most of our relationships—in business, in our neighborhoods, even with our closest friends. Being first, no matter how, has permeated our lexicon as a given: *All's fair in love and war. Survival of the fittest. Winner take all. He who dies with the most toys wins.* It is hardly surprising that highly competitive tactics have crept into the social relations of our children, leading to transgressions, large and small.

I began to think about the social exchange in my own neighborhood, and about how much of what psychologists call "relativity awareness" has played a part. *How many children do you have? What kind of car do you drive? How many vacations are you off on this year?*

Which college has your kid gotten into? What's his or her grade point average? Where, in other words, do you fit on the social ladder? Even the best of us have internal moments like *American Psycho*'s Wall Street trader Patrick Bateman when regarding his colleague's exquisite new business card with alarm. "Oh my God," he thinks. "It even has a *watermark.*"

Nevertheless, the notion that competition is a fundamental human urge made no sense to me from a scientific perspective. I write about frontier science, and the latest evidence from many disciplines—from neuroscience and biology to quantum physics—suggests that nature's most basic drive is not competition, as classic evolutionary theory maintains, but wholeness. I'd seen a good deal of new research demonstrating that all living beings, including human beings, have been hardwired to seek connection above virtually any other impulse—even at personal cost. Nevertheless, our current paradigm, as provided us by traditional science, maintains a view of the universe as a place of scarcity populated by separate things that must turn against each other in order to survive. For me to win, you have to lose. We've all simply assumed *that's life.*

Furthermore, it might have been our view of life, but it certainly wasn't true of the one animal closest to hand, our dog, Ollie. There is nothing dog-eat-dog about Ollie's neighborhood. He doesn't have much time for humans, but he is kind to every dog he meets on his walks. He regularly shoves bones under the fence for T-bone, our next-door neighbor's affenpinscher; in fact, he saves the biggest bones for her. Ollie's relationship with T-bone defies every current biological description about the imperative to act selfishly. Since T-bone is neutered, there's no personal genetic advantage to courting T-bone, no possibility of extending the family line. Nevertheless, when T-bone visits, Ollie will raid our garbage wheelie bin, fishing out a chicken carcass for her to feast on, then offer her unfettered access to his own food bowl, his pigs' ears, and his toys. Although T-bone is smaller than he is, when Ollie plays with her, he often lets her win just to keep her in the game.

I began to ask myself a basic question: *Does it have to be like this?* Were we meant to be so competitive with one another? Is it inherent in animal and human biology? How did it get like this? And if we're not this, what are we supposed to be?

Since that dress rehearsal, I've been thinking that, at some point, we'd torn up the social contract and forgotten how to come together. Somewhere along the line, we'd forgotten how to *be*.

It doesn't have to be like this. As I began researching this book and studied the latest discoveries in a vast array of disciplines—general biology, physics, zoology, psychology, botany, anthropology, astronomy, chronobiology, and cultural history—the more it became clear to me that the lives we've chosen to lead are not consistent with who we really are. I discovered other societies that live very differently from us, with a world view more in keeping with the findings of the new science. These cultures conceive of the universe as an indivisible whole, and this central belief has bred an extraordinarily different way of seeing and interacting with the world. They believe that they are in relationship with all of life—even with the earth itself. We see the thing; they see the glue between the things—the thing that holds them together. The essential thing for these societies is not the individual, but the relationship between individuals, which they view as a thing in itself.

They've understood the essential nature of humanity as a coming together—a communion—and as a consequence, they live happier lives with lower divorce statistics, fewer troubled children, less crime and violence, and a stronger community.

They have chosen a better way to live, a more authentic way to be—the way, I believe, that you and I were meant to live. And they do so because they've bought into another narrative—another world view of who we are and why we're here than that espoused by our culture, most particularly by our current science.

I wrote this book to prove to you that we are operating according to an outdated set of rules. I want to show you that the scientific story of who we are has drastically changed, and that we must change with it in order to survive. The competitive impulse that is now a major part of our self-definition and that forms the undercurrent of all our lives is the same mindset that has created every one of the large global crises now threatening to destroy us. If we can recover wholeness in our relationships, in my view, we will begin to heal our world.

My hope is that reading *The Bond* will give you a better neighborhood than the one you have now—so that constant comparison and stealing roles aren't part of the rules of the game.

It should be pointed out that this book is emphatically not an apology for communism or socialism, nor is it a new economic or political model. I enjoy the freedoms of democracy, just as I suppose you do, if not the recent excesses of the economic model it has spawned, and after living in the UK for many years, I have witnessed firsthand the shortcomings of certain of its older, more socialist policies. The ideas in this book are not meant to disparage either science as a profession or the discoveries of the great geniuses of science like Isaac Newton or Charles Darwin. I am emphatically not a creationist. Nevertheless, science is a relentless process of discovery. No one scientist writes the final story for all time. As new findings come to light, new chapters constantly revise and even supplant earlier versions. Currently, we are undergoing a radical revision of our understanding of ourselves and our world. Many of the theories we have held sacred, including the original theory of evolution, are being refined as more information about the nature of our world unfolds.

I hope to offer you the first roadmap of how to live according to this new scientific story. I aim to give you a fresh way of seeing the world from a wider perspective, a new way of relating to other human beings, a new sense of community, a new and authentic purpose—something better than envying your neighbor's watermarked business card.

I intend to demonstrate to you how easy it is to live in wholeness, how tiny changes can revolutionize your life and the life of everyone around you.

It doesn't have to be like this. Not for one more day.

Lynne McTaggart,
December 2010

INTRODUCTION

We sense that we have reached the end of something. Since the millennium, commentators of every variety have been trying to get a handle on the collective significance of the continuous crises besetting us in modern times: banking crises, terrorist crises, sovereign debt crises, climate change crises, energy crises, food crises, ecological crises, manmade and otherwise.

"The world as we know it is going down," a Wall Street broker told reporters in September 2008, after Lehman Brothers collapsed and Morgan Stanley threatened to follow suit. It is the "end of capitalism as we know it," declared the filmmaker Michael Moore when the American auto giant General Motors filed for bankruptcy. It is the end of our dependence on fossil fuel, announced President Barack Obama, about the Deepwater Horizon oil rig explosion. It is the end of nature, wrote Bill McKibben in his book of the same name. It is the end of oil, wrote the journalist Paul Roberts in *his* book of the same name. It is the end of food because it is the end of oil, declared Roberts in his follow-up book. For those who take stock in the Mayan Long Count calendar and the apocalyptic significance of 2012, it is the beginning of the end of the world.

But the crises we face on many fronts are symptomatic of a deeper problem, with more potential repercussions than those of any single cataclysmic event. They are simply a measure of the vast disparity between our definition of ourselves and our truest essence. For hundreds of years we have acted against nature by ignoring our essential connectedness and defining ourselves as separate from our world. We've reached the point where we can no longer live according to this false view of who we really are. What's ending is the story we've been told up until now about who we are and how we're supposed to live—and in this ending lies the only path to a better future.

In this book, I have an audacious mission: to revolutionize the way you live your life. This book is going to rewrite the scientific story you've been told about who you are, because the current version has reduced us to our lowest common denominator. At this very moment you live contrary to your truest nature. I hope to help you to recapture your birthright, which has been sabotaged not only by modern society but, more fundamentally, by modern science. I wish to wake you up to who you really are, to do nothing less than to return to you your authentic self.

The leitmotif of our present story is the hero up against it all. We take it for granted that our life's journey is meant to be a struggle. Consequently we remain constantly vigilant, poised to wrestle with every behemoth—at home, at work, among our acquaintances and friends—that strays across our path. No matter how pleasant our lives, the vast majority of us maintain a stance of operating *contra mundi*, with every encounter some sort of battle to be fought: against the coworkers who seek to usurp our jobs or promotions, or the students who raise the bell curve against which we are judged; against the people who take our subway seat, the shops that overcharge us, the neighbors who have a Mercedes when we drive a Volvo, and even the husband or wife who has the temerity to insist on maintaining an opinion that is different from ours.

This idea that we operate *against* the world has its origin in our basic understanding that this self of ours, the thing we call *I*, exists as a separate entity, a unique creation of genetic code that lives apart from everything else out there.

The most enduring statement we make about the human condition, the central fact of our existence, is our solitude, our sense of separation from the world. We regard as self-evident that we exist as self-contained, isolated beings, living out our individual dramas, while everything else—other atoms and other cells, other living things, the land masses, the planets, even the air we breathe—exists as something distinct and wholly separate.

Although we begin life from the uniting of two entities, from there on in, science tells us, we are essentially on our own. The world is the irrefutable *other,* carrying on impassively with or without us. Our hearts, we believe, beat finally and painfully alone.

This paradigm of competitive individualism offers us a view of life as a heroic struggle for dominion over hostile elements and a share of strictly limited resources. There's not enough out there, and others may be fitter than we, so we have to do our damnedest to get hold of it first.

A multitude of influences—religious, political, economic, scientific, and philosophical—writes the story that we live by. Nevertheless most of the big ideas we have about the universe and what it is to be human derive from three revolutions: the Scientific Revolution, or the Age of Enlightenment, and the two Industrial Revolutions of the eighteenth and nineteenth centuries, which transformed the cultural and socio-economic conditions of the West into our modern developed world. These movements largely created the modern sense of our own individuality by drastically altering our vision of the universe from a harmonious, benevolent, and interconnected whole to an amalgam of separate and unrelated things, competing with each other for survival.

The Scientific Revolution launched a relentless march toward atomization, as scientists believed they could understand the whole of the universe by studying its individual components.

With the publication of the *Philosophiae Naturalis Principia Mathematica* in 1687, Isaac Newton, the father of modern physics, described a universe in which all matter was thought to move according to certain fixed laws within three-dimensional time and geometrical space. Newton's laws of motion and gravity depict the universe essentially as machine, a vast clockwork of separate parts that can always be relied upon to follow predictable behavior. Once Newtonian laws demonstrated that the trajectory of virtually everything, from single objects to the motion of the planets, could be reduced to a mathematical equation, the world came to be viewed as dependably mechanistic. Newtonian laws also demonstrated that things exist independently from each other, complete in themselves, with their own inviolate boundaries. We ended with the hairs on our skin, at which point the rest of the universe began.

The French philosopher René Descartes wrote of man's essential separation from his universe in a philosophy that banished any kind of holistic intelligence from nature and portrayed matter as mechanistic and corpuscular. Even our material bodies lay outside of our conscious selves: one more well-oiled and highly dependable machine.

The Newtonian paradigm of world-as-machine was further reinforced with the arrival of the most influential machine of all: the steam engine. Steam and the development of machine tools not only transformed the production of food, fuel, heating, manufacturing, and transport; they also profoundly affected human beings by separating them from the natural world. In every way life was broken down into regular sequences. Work was now dictated by an assembly line, and workers became one more cog in the wheel of production. Time was parceled out in minutes, and not through the seasons of planting and harvest, and marked through the punching in of a clock. The vast majority of people working in factories no longer followed the rhythms of nature, but the rhythms of the machine.

The Second Industrial Revolution, in the nineteenth century, introduced modern technology with the advent of steel and petroleum manufacturing and led to the rise of a middle class, which in turn paved the way for modern capitalism and the promotion of the individual and his interests. In *An Inquiry into the Nature and Causes of the Wealth of Nations,* published in 1776 and considered one of the founding philosophies of economic theory, Scottish philosopher Adam Smith argued that the "invisible hand" of the market, created by natural supply and demand, and competition between self-interested individuals would naturally best serve society as a whole. He famously believed that we do best for others by giving way to this fundamentally selfish nature of ours and looking out for Number One: "By pursuing his own interest, [the individual] frequently promotes that of the society more effectually than when he intends to promote it."[1]

Undoubtedly the scientific discovery with the most pervasive hand in our current worldview is Charles Darwin's theory of evolution through natural selection. When assembling his ideas for *On the Origin of Species,* the young Darwin was profoundly influenced by the concerns of the Reverend Thomas Robert Malthus on population explosion and limited natural resources.[2] Darwin concluded that, since there wasn't enough to go around, life must evolve through what he termed a "struggle for existence." "As more individuals are produced than can possibly survive," Darwin wrote in *Origin,* "there must in every case be a struggle for existence, either one individual with another of the same species, or with the individuals of distinct species, or with the physical conditions of life."[3]

Darwin was at pains to note that his catch phrase "struggle for existence" was not literal but highly elastic, encompassing everything from the search of tree roots for water to the reliance of a pack of animals on each other. It was actually the British philosopher Herbert Spencer who first coined the term "survival of the fittest," after an enthusiastic reading of *On the Origin of Species;* after some persuasion Darwin accepted the term,[4] eventually adding the subtitle: *Or The Preservation of the Favoured Races in the Struggle for Life.*

Malthus had provided Darwin with a central metaphor to explain the mechanism behind nature's drive to propagate and thrive, and as an inadvertent consequence, Darwin unleashed upon the world a metaphor that came to represent the human experience: *life as war.* An individual or population thrives only at another one's expense. *I win, you lose.* Despite Darwin's liberal use of the term, almost immediately the narrower meaning of the metaphor stuck, offering a scientific framework for all the various burgeoning social and economic movements of the day. Most subsequent interpretations of Darwin's work, even in his lifetime, promoted a vision of all aspects of life as a battle over scarce resources, in which only the toughest and most single-minded survived.

The English biologist Thomas Huxley, the Richard Dawkins of his day, dubbed "Darwin's bulldog" for his role as Darwin's vociferous mouthpiece, generously extended the view of dog-eat-dog competition in his belief that it was responsible for the evolution of culture, ideas, and even the human mind. Huxley was convinced that it was in the natural order for human beings to put their own interests above all others.[5]

Thanks to newly invented telegraphic cables and advances in printmaking, the wider interpretation of Darwin's theory quickly swept across the globe. "Survival of the fittest" made for a perfect fit with Smith's brand of enlightened competition in the marketplace, but besides Western capitalism, the theory of natural selection was also used to justify the Chinese revolution and the "whitening" of Latin American indigenous culture with European stock.[6] Writers such as the Russian-born Ayn Rand used fiction as thinly disguised polemic to applaud the process whereby each of us attempts to gulp the biggest breath of a strictly limited amount of oxygen.

The metaphoric representation of life as a race to the finish line has been used as intellectual justification for most aspects of modern industrialized society, which regards competition as society's perfect shakedown mechanism, separating out the economically, politically, and socially weak from the strong. The winners have a right to winner take all because the human race as a whole would benefit from it.

The final important influence on our modern scientific definition of ourselves occurred in 1953, when the molecular biologists James Watson and Francis Crick claimed to have unlocked the "secret of life" by unraveling deoxyribonucleic acid (DNA), the genetic coding in the nucleus of every cell. Thereafter many scientists came to believe that within the coiled double helix lay every individual's lifelong blueprint. Each of our cells, equipped with a full pack of genes, would live out its preprogrammed future, while we were held hostage by our genetic destiny, powerless to do anything other than observe the drama unfolding. As with every other kind of matter, the human being had also been atomized—reduced, in a sense, to a mathematical equation.

Modern-day interpreters of Darwin, the neo-Darwinists, have woven competition and struggle into the latest theories of our biological makeup by proposing that every part of us acts selfishly in order to survive; our genes—even our ideas—are engaged in competition with other gene pools and thoughts for domination and longevity.[7] Indeed some scientists invest genes with the power to control every aspect of our lives, considering the body an accidental byproduct of a greater evolutionary endeavor.[8]

Modern evolutionary theory has removed any vestigial sense of moral design or beneficence from the natural world: nature has no stake in cooperation or partnership, but only likes winners, of any sort. The vision of a purposeful and harmonious whole has been replaced with blind evolutionary force, in which human beings no longer play a conscious part.

Many psychologists argue that competitiveness is hardwired within us, a natural biological urge as inherent as our basic urge to survive. After we stop fighting over food, water, shelter, and mates, the theory goes, we begin competing over more ephemeral prizes: power, status, and, most recently, fame.

Consequently for more than three hundred years our worldview has been shaped by a story that describes isolated beings competing for survival on a lonely planet in an indifferent universe. Life as defined by modern science is essentially predatory, self-serving, and solitary.

These metaphors—the mechanistic view of the universe, the "red in tooth and claw" sense of ourselves—have seeped into our consciousness to permeate our every day. Our paradigm for living today has been built upon the premise that competition is the essential calling card of existence. Every modern recipe in our lives has been drawn from our interpretation of life as individual and solitary struggle, with every-man-for-himself competition an inherent part of the business of living. Our entire Western economic model is built on the notion that competition in a free-market economy is essential to drive excellence and prosperity. In our relationships we extol our inherent right to individual happiness and self-expression above all else. We educate our young by encouraging them to compete and excel over their peers. The currency of most modern two-cars-in-every-garage neighborhoods is comparison and one-upmanship. The world, as Woody Allen once put it, "is one big cafeteria."

The individualistic, winner-take-all zeitgeist of modern times is to blame for many of the crises we presently face in our society, particularly the excesses of the financial sector, with its insistence on a bigger and better profit every year, at any cost. Before being jailed for his part in the energy company Enron's vast array of fraudulent activities, CEO Jeffrey Skilling bragged that his favorite book was neo-Darwinist Richard Dawkins's *The Selfish Gene,* and he periodically fired the entirety of those of his workforce achieving the bottom 10 percent results as a means of improving the overall "fitness" of the herd.[9] This mind-set is responsible for the raft of deceit that now goes on in every sector of society, from the 50 percent of college students known to cheat on exams to corporate cheating, even in sectors designed for the public interest. Up to three-quarters of all published research on pharmaceutical drugs in the medical literature is now believed to be ghostwritten by public relations firms hired by drug companies, with serious and even potentially fatal side effects routinely concealed.[10]

The danger inherent in our current worldview is apparent in the ways that it has been taken to its extreme conclusion and used as a justifying principle for sociopathic behavior, from mass murder during the Third Reich and the eugenics of the twentieth century, to modern ethnic cleansing and serial murder. Eric Harris, for instance, was sporting a T-shirt emblazoned with the words "Natural Selection" on April 20, 1999, when he and Dylan Klebolt, armed with propane bombs, assorted Molotov cocktails, an Intratec TEC-DC9 blowback-operated semiautomatic handgun, a Hi-point 995 Carbine semiautomatic pistol, a Savage 67-H pump-action shotgun, and a Stevens 311D double-barrel sawed-off shotgun, sauntered into Columbine High School and opened fire.[11]

Although our Newtonian vantage point has afforded us technological mastery over our lives, the worldwide collapse of our global economic model in 2008, the current ecological crises, the threatened shortages of water and food, the exhaustion of petroleum sources all expose the extreme limitations of this mind-set, which now threatens our planet with extinction. On a personal level it has left most of us with a distinctly hollow feeling, as though something profound—our very humanity—has been trampled on in our daily wrestle with the world.

We urgently need a new story to live by.

For the past several decades, ever since I began pondering the meaning of cutting-edge discoveries in physics and other branches of science,[12] I have been struck by how much of scientific theory, and consequently our model of the way things work, is currently going up in smoke. With every new finding in the sciences, yet another cherished notion we hold about ourselves is overturned. An entirely new scientific story is emerging that challenges many of our Newtonian and Darwinian assumptions, including our most basic premise: the sense of things as separate entities in competition for survival. The latest evidence from quantum physics offers the extraordinary possibility that all of life exists in a dynamic relationship of cooperation. Quantum physicists now recognize that the universe is not a collection of separate things jostling around in empty space. All matter exists in a vast quantum web of connection, and a living thing at its most elemental is an energy system involved in a constant transfer of information with its environment. Rather than a cluster of individual, self-contained atoms and molecules, objects and living beings are now more properly understood as dynamic and protean *processes,* in which parts of one thing and parts of another continuously trade places.

This revolution is not confined to physics. Extraordinary new discoveries in biology and the social sciences have profoundly altered our view of the relationship between living things and their environment. Frontier biologists, psychologists, and sociologists have all found evidence that individuals are far less individual than we thought they were. Between the smallest particles of our being, between our body and our environment, between ourselves and all of the people with whom we are in contact, between every member of every societal cluster, there is a *Bond*—a connection so integral and profound that there is no longer a clear demarcation between the end of one thing and the beginning of another. The world essentially operates, not through the activity of individual things, but in the connection between them—in a sense, *in the space between things.*

The most essential aspect of life is not the isolated thing, whether a subatomic particle or full-fledged living being. It is the relationship itself: the inseparable, irreducible *Bond.* This connection—the space in between—holds the key to the life of every organism, from subatomic particles to large-scale societies, and also the key to our viable future.

These discoveries suggest that the idea of the individual as an individual thing distinct from other things is a fallacy. There is nothing—from our subatomic molecules to our entire being—that we can define with any certainty as a wholly separate body that can be isolated and ring-fenced. The "individual" is only the sum of an infinite number of inexactly defined parts, and the parts as we currently understand them are shifting and transforming at every moment. In every way individual things live life inextricably attached and bonded to an "other." Nature's most basic impulse is not a struggle for dominion but a constant and irrepressible drive for wholeness.

These new revelations from the frontiers of science represent a reversal of the process of atomization that began with the Age of Enlightenment. The new story that is being written around the globe adds up to the beginning of a recovery of our holistic view of ourselves as inextricably bound to everything we see around us.

These discoveries hold not only vast implications about how we choose to define ourselves, but also vast implications about how we ought to live our lives. They suggest that all our societal creations, invested as they are in competition and individuality, run counter to our most fundamental being—that a drive for cooperation and partnership, not dominance, is fundamental to the physics of life and the biological makeup of all living things. They imply that most of us in the developed world are not living in harmony with our true nature. That we are constantly affecting and being affected by all matter in a constant and ever evolving Bond demands a drastic change in the way we relate to ourselves and all other living things.

We need some new rules to live by. We need another way to *be*.

This book offers a new metaphor to live by. It rejects the central tenet of orthodox science: that matter, even subatomic matter, exists in isolation and is complete in itself. It suggests that the dance of life is not a solo, but a duet—that every part of you connects to an essential and irreducible *Bond*. It acknowledges that each of us is so tightly interconnected with our world that we can only hope to live authentically when we live by a very different story.

We need to adopt a new definition of what it means to be human. We need to look at our universe with a fresh pair of eyes. Applying these new discoveries to every aspect of our lives requires nothing less than making ourselves anew.

The Bond ultimately posits an alternative future in which a new paradigm for living in partnership and connection replaces the metaphor of battle. I hope to offer you an entirely new vision of yourself and your place in the world, not as its master or its competitor, but as its cooperative partner.

That new vision starts with the understanding—shocking in the breadth of its implications—that nothing in the world is separate. In fact in the most basic sense there is no such thing as a thing.

PART I

THE SUPERORGANISM

The sickness of our times for me has been just this damn thing that everything has been getting smaller and smaller and less and less important.

Norman Mailer, *The Naked and the Dead*

CHAPTER 1

THE HUNT FOR THE THING

On a bench the size of two SUVs at the University of California at Berkeley, Graham Fleming and his colleagues in the chemistry department have set up the scientific equivalent of a pinball machine. Numerous precision lasers, which pulse out light at millions of a billionth of a second, are placed at various strategic points and trained onto an obstacle course of mirrors and glass lenses, themselves aimed at a tiny solitary black box. Once the machines are switched on, the laser light generated by these ultrafast devices will career off each mirror and lens before shooting inside and alighting on the box's contents: a tiny sampling of a green sulfur bacteria. The light from the lasers is supposed to mimic the sun, for this type of bacteria, for all intents and purposes, is a plant, with the same extraordinary photosynthetic ability to convert sunlight into energy inside its cells.

By tracking the means by which a rudimentary living thing harnesses the power of the sun and converts it into stored energy, Fleming, a British-born sixty-year-old, hopes to solve the central mystery of plants: their ruthless efficiency. The miracle is not only that the plant can manage this feat at all, but that it does so by using every last photon that comes its way.

The most sophisticated machine on earth cannot begin to mimic the energy production of a plant. Every manmade activity of rough equivalence diminishes the initial store of energy by more than 20 percent in the process of transforming from one type of energy to another. If humans could learn to capture and transform solar energy through even a crude approximation of the manner in which plants do, mankind's future energy needs would be forever secure.

The other aspect of the mystery is more elementary: how a simple living system like a plant can generate the world's oxygen and carbohy-

3

drates through a reaction powered by the electricity that it essentially creates from light.

The key to studying this extraordinary process lies in tracking the path of electron energy inside the protein scaffolding of the cell, which connects the bacteria's exterior solar panels, or chlorosomes, the harvesters of sunlight, to reaction centers at the heart of the cells, the tiny crucible where the miracle of conversion takes place.

Fleming's experiment takes a tiny fraction of the time it takes to flicker an eyelid. As soon as the pulsed light from the lasers hits the protein, it excites the electrons, and the resulting energy then needs to find the most direct route along the tiny protein scaffolding track to the reaction centers. This is a complex and potentially time-consuming task, according to conventional physics, as there are many possible pathways and end points that the electron's energy must seek out and eliminate, one by one.

What Fleming discovered is nothing less than a giant chink in the entire edifice of accepted biology. Rather than a single pathway, the energy reaches its target by trying out several routes simultaneously. Only when the final connection is made and the end of the road reached does the energy track its most efficient course—retroactively—and the energy follow that single path. It appears as if the optimum route were chosen *backward in time*, after all possibilities have been exhausted. It is as if a person lost in a labyrinth had tried all possible pathways at the same time, and after finally discovering the correct pathway to the exit, eliminated all trace of his rehearsals.

Fleming's discovery is a wholly unexpected answer to his line of inquiry: the plant is so efficient because the energy generated by its messenger electrons is able to occupy more than one location at the same time.

Fleming is making some of the first tentative forays into what has been called "quantum biology," producing the first evidence that life on earth is driven by the laws of quantum physics, and his experiment is necessarily crude. It substitutes laser light for true sunlight and is carried out at temperatures of 70 kelvin (or −333.7°F), an environment far too cold for most plants to survive.

Nevertheless, with his background in physics as well as chemistry, Fleming realizes the import of what he has just witnessed. As the founders of quantum theory, the Danish physicist Niels Bohr and his brilliant German protégé, Werner Heisenberg, discovered in the early part of the twentieth century, subatomic particles like electrons and photons by themselves aren't an actual *anything* yet. Atoms are not little solar systems of billiard balls but rather a messy little cloud of probability. They exist in many places simultaneously, in a state of *pure potential*—or, as physicists refer to it, "superposition"—the sum of all probabilities. A subatomic particle like those in Fleming's bacteria essentially experiments with this pathway and that pathway at the same time before choosing the optimum pathway to the reaction site.

One of the conclusions of their theory, which has become known as the Copenhagen Interpretation, after the city where Bohr and Heisenberg first hammered out the inescapable conclusions of their mathematical discoveries, is the idea of *indeterminacy,* the fact that you can never fully know everything about a subatomic particle. If you measure where it is, for instance, you cannot also work out where it is going or at what speed. Bohr and Heisenberg recognized that a quantum particle can exist as both a particle, a congealed, bullet-like thing, and a "wave function," a big smeared-out region of space and time, any corner of which the particle may occupy.

In a quantum state a particle exists as a collection of all possible future selves all at the same time, like an endlessly replicated chain of paper dolls. An electron "probably" exists until scientists pin it down and take a measurement, at which point its multiple selves collapse and the electron settles down into a single state of being.

If the results of Fleming's experiment are verified—and others have now successfully carried out the experiment on real plants at room temperature—this will mean that the most fundamental process of the universe, the process responsible for life on earth, is driven by a mechanism that isn't actually anything at all, at least according to our usual definition of things. The electron driving photosynthesis is a will-o'-the-wisp, impossible to pin down or locate with precision.[1] Fleming's experiment also lays bare a much larger possibility: that all of life is created and sustained by something so ephemeral that we may not even be able to identify what it actually is, much less locate where it is with precision.

Although revolutionary in its implications, Graham Fleming's discovery is not especially revelatory to quantum physicists. Many within this discipline have been casting about without success to find *the thing:*

the smallest thing that creates all other things in the world. All modern suppositions about our physical universe rest on the belief that life is composed of things, which in turn are made up of littler things, and that we can understand the big things by seeking out and naming the little things.

Ever since a Muslim physicist named Ibn al-Haytham developed the scientific method more than a thousand years ago, scientists have attempted to take apart the universe like one vast radio to examine its component parts. For the past hundred years or so they have been preoccupied with attempting to locate the tiniest of its building blocks. In 1909 the Nobel Prize–winning New Zealand chemist Ernest Rutherford and his colleagues at the University of Manchester created the Rutherford model of the atom, a tiny solar system of orderly electrons, after discovering what at first was believed to be its sun and one of the world's smallest units: the nucleus. Rutherford's model took a slight hammering when a colleague from Cambridge, the British physicist James Chadwick, went on to discover an even smaller particle inside the nucleus: the neutron.

Chadwick posited that the constituents of an atom, the protons, electrons, and neutrons, are the most fundamental units of our world—until it was discovered that, like a Russian doll, within these particles lay still smaller particles.

In 1969 science briefly congratulated itself on isolating what it thought was the most essential of the universe's elements when the quark was discovered—until an alphabet soup of other particles was found or postulated in the following decades: muons and tauons, positrons and gravitons, particles with force and particles without force, upsilon particles, tau neutrino, and the most recent discoveries, skyrmions and goldstinos and dyons and pomerons and luxons, plus strongly interacting "composite particles" like hadrons and even hypothetical particles born of supersymmetry theories.

To make sense of all these entities, physicists produced the Standard Model, the Rosetta Stone of modern particle physics, which lumps all these hundreds of varieties of particles and impossibly complicated interactions into three families and their fundamental interactions and flavors: six types of quarks, six leptons, and a variety of bosons, or "force carrier" particles, which include the tiniest unit of light, the photon; gluons, something called weak-gauge bosons, plus gravitons and the Higgs boson, the latter two classifications assumed to exist but never actually yet seen.

However elegant the Standard Model is as a theory, enabling scientists to reduce all these dozens of particles into mathematical shorthand, the bottom line is that physicists still cannot isolate one single structure and claim with any confidence that this is it, the smallest currency of the universe, the final individual entity out of which our world derives. Most of the dozens of particles discovered after World War II are now thought not to be elementary but rather composites of particles; in fact physicists now allow that it may be impossible ever to prove that these particles can be further separated into their component parts.

Physicists *assume* that certain particles are more elementary than others—that quarks are more elemental than, say, nucleons or pions. Nevertheless, as the Nobel Prize–winning American particle physicist Steven Weinberg once lamented, "We cannot reach any final conclusion about the elementarity of the quarks and gluons themselves."[2]

What scientists have settled for, in the Standard Model theory, is a fuzzy approximation that may have as much to do with the final truth of life as a cyborg has to do with a human being. The Standard Model is likely to prove only a vague approximation for some more fundamental theory that will reveal itself once scientists have invented higher energy particle accelerators, at which point we might discover that the tiniest of these particles isn't in fact the smallest Russian doll but simply another doll with more dolls inside.

One reason for this continuing difficulty in locating the smallest piece of the universe may be the simple fact that nothing, finally, exists independently. Although we consider matter discrete and definable, the fact is that it cannot be compartmentalized into anything definitive. Even the smallest structure of matter may prove impossible to separate from its neighbors, place a fence around, and say with any finality that here is where it begins and there is where it ends. For anything smaller than an atom we cannot figure out if a subatomic thing exists on its own or as a composite of elements.

The closer scientists look, the more they discover how dependent on, and finally indivisible from, everything is with everything else. Werner Heisenberg referred to this fact as the "most important experimental discovery of the last 50 years." He also noted that even the question of what particles "consist [of] no longer has any rational meaning": "A proton, for example, could be made up of neutron and pion, or Lambda-hyperon and kaon, or out of two nucleons and an anti-nucleon; it would be simplest of all to say that a proton just consists of continuous matter, and all these statements are equally correct or equally false. *The difference between elementary and composite particles has thus basically disappeared.*"[3] In fact the very word "particle," with its suggestion of a separate and corporeal reality, is a misnomer. When particle physicists get down to the bottom layer of matter there isn't really anything there. Although Rutherford's model is still taught in high school physics, and atoms are still rendered as a grouping of tiny, well-behaved billiard balls with orderly little orbits around a central nucleus, subatomic particles more closely resemble a tiny coalescence of energy, a smeared-out, uncongealed puff of vibratory nothingness.

Vlatko Vedral, a professor of quantum physics at Oxford University, once remarked that it is more correct to say that a particle is an excitation of a wave, an excitation of energy, a little coalescence of energy within a larger field of energy, much as a knot exists on a length of rope. Steven Weinberg adds, "We will not be able to give a final answer to the question of which particles are elementary until we have a final theory of force and matter. When we have such a theory, we may find that the elementary structures of physics are not particles at all."[4]

Although we classify everything in the universe as separate and individual, individuality, at the most rudimentary level, does not exist.

The world owes a debt of gratitude to Werner Heisenberg's delicate immune system and his body's tendency toward excessively high histamine levels. During a severe attack of hay fever in May 1925 Heisenberg packed himself off to the main island of Heligoland, a small, rocky archipelago off the northwest coast of Germany, whose name, "Holy Land," is a nod to its unusually temperate climate—a meteorological freak in the North Sea—and the virtual absence of pollen on its inhospitable terrain. Once he began to breathe easily, Heisenberg could also freely contemplate many of the conundrums posed by new discoveries about the quantum structure of matter. He and Niels Bohr had held many long discussions about the incompatibility between these new ideas and existing physics theory while hiking in the mountains near Bohr's home in Copenhagen, and Heisenberg had spent long hours trying to reconcile the two. With no other distraction on that treeless island, he finally worked out the beautifully elegant equations of quantum matrix mechanics, ending forever all need to shoehorn the new discoveries about quantum entities into classical mechanics.

Heisenberg's ideas reduced down to a simple truism: any theory about the physical universe should be concerned only with what can actually be observed in experiments. He'd stripped away all the suppositions that had been made about subatomic bits, such as any notion that they orbit like planets around the sun. Rather than working individual digits, he played around with a cluster of numbers to represent the spectrum of states open to a subatomic entity, and in this manner ultimately found a mathematical means to represent the strange twilight of the quantum particle.

After returning to the mainland Heisenberg showed his work to Bohr and his other mentor, the physicist Max Born, who helped to formulate it into the first consistent theory of quantum physics.

Heisenberg had been left with one strange fact with his equations, which were otherwise so successful: they did not commute. Unlike in normal algebra, $x + y$ did not equal $y + x$. A year later he went on to formulate the Uncertainty Principle, or *Ungenauigkeit* in his native German, which essentially put forward the outrageous proposition that matter is finally unknowable. By confining himself to the purity of mathematics, he'd discovered a way to demonstrate what was beginning to dawn on both him and Bohr. They'd discovered something so counterintuitive and bizarre about the physical world that many modern physicists, imbued with Newtonian physics, continue to have difficulty accepting it: there is no such thing as a thing, no underlying solid reality, but only *the space between,* the indivisible relationships between things.

Since the establishment of quantum physics physicists have felt compelled to keep inventing new theories—string theory, many-worlds theories—because otherwise, although the math works, the logic of it continues to confound. Nevertheless modern quantum physicists have demonstrated what Heisenberg had first intuited: matter is nothing but a relationship; $x + y$, in a sense, stands for an impenetrable bond between two indeterminate things that do not exist on their own. Or as Heisenberg, with his philosophical bent, would bluntly note, after fruitlessly attempting to move beyond uncertainty in the quantum world, "We cannot know, as a matter of principle, the present in all its details."[5]

Heisenberg went on to refine his theories in what became known as "quantum field theory." He discovered that at our most fundamental layer of being, our subatomic particles not only aren't really a definable anything, but also do not remain the same at any moment. The little things that make up big things in the universe don't stay the same, but keep changing. All subatomic particles are constantly trading information with their environment and being reshuffled in a dynamic pattern. The universe contains an indeterminate number of vibrating packets of energy that constantly pass energy back and forth as if in an endless game of basketball with a quantum sea of light. Indeed they aren't even there all the time, but are constantly popping in and out of existence, making a brief appearance before disappearing back into the underlying energy field.

All elementary particles interact through what are considered temporary or "virtual" quantum particles, combining and annihilating each other in less than an instant. Furthermore every type of particle has its shadow self in the form of antimatter or an antiparticle, which behaves just like its corresponding "positive" variety except with an opposite charge. So for every quark there is an antiquark, for every electron a positron. And should the two ever meet they simply combine, so that the superficial appearance of an entity reverts to indeterminate, unspecified energy.

The back-and-forth passes of virtual particles, akin to two people taking turns constantly depositing and withdrawing the same amount of money from a bank, are known collectively as the *Zero-Point Field*. The field is called "zero point" because even at temperatures of absolute zero, when all matter theoretically should stop moving, these tiny fluctuations are still detectable. Even at the coldest place in the universe, subatomic matter never comes to rest, but carries on this little energy tango.[6]

Nature's most basic ingredients are bundles of energy that are indistinguishable from the field around it. According to quantum field theory, the individual entity is transient and insubstantial, and particles cannot be separated from the empty space around them. Although you appear the same at any given moment, you are an entirely new batch of subatomic energy with every breath you take.

Rather than a batch of separate things jostling around in empty space, it is more correct to say that fundamental matter is simply a relationship between two indeterminate things: particle energy traded with other particle energy and also with the background Field. It is in fact the Bond between these tiny particles and the background Field that creates everything that we refer to as "matter." All matter depends on a connection with this most fundamental of energy fields to achieve a sense of solid and stable substance.

Hal E. Puthoff, the director of the Institute for Advanced Studies at Austin in Austin, Texas, and his colleagues have shown that this constant to-and-fro exchange of all subatomic matter with the Zero-Point Field accounts for the stability of the hydrogen atom, which implies that it also accounts for the stability of all other matter.[7] Without this constant trading of energy with the Zero-Point Field, the electrons within the atoms of things would spiral out of control and be sent crashing into the atom's nucleus, and all matter would simply implode.

Puthoff also showed that this relationship creates the sense we have that an object has mass, or density.[8] In a landmark paper he and his colleagues argued that inertia—the tendency of objects to stay put and be hard to get moving, and then once moving, be hard to stop—is simply a "drag force," or resistance to being accelerated through the Zero-Point Field. The larger the object, the more particles it contains and the more tightly it is gripped onto by the Field. The "drag" on that object gives the appearance of mass, but in the eyes of these physicists, mass is simply energy holding onto other energy. Whenever you push on an object or it attempts to move, the interaction between the particles, or energy vibration, of that mass is held onto tightly by the Zero-Point Field, giving the illusory sense of corporeal form.

What this essentially boils down to is that everything we label an object, no matter how large or how heavy, is essentially a collection of electric charges interacting with other energy. The most basic property of matter, its sense of being a solid "something," is only and entirely due to the *Bond* between subatomic particles and the background sea of energy.[9] A subatomic "particle" is simply the seeking of a connection in the space between a big web of energy and a little knot of energy. You and everything around you are simply a collection of charged energy having a relationship.

We also now understand that these little knots of energy mostly like to operate as an indivisible collective. Another strange feature of quantum physics is *nonlocality*, also referred to as "entanglement," with its poetic connotation of doomed indivisibility, like a pair of star-crossed lovers who may be forced to separate but remain forever mentally and emotionally intertwined. Niels Bohr discovered that, once subatomic particles such as electrons or photons are in contact, they remain aware of and forever influenced by each other instantaneously and for no apparent reason, over any time and any distance, despite the absence of force or energy, the usual things that physicists understand are necessary for one thing to affect something else.

When particles are entangled the actions of one will always influence the other in the same or the opposite direction, no matter how far apart they are separated, like Siamese twins that can never be divided. Once they have connected, the measurement of one subatomic particle instantaneously affects the position of the second one. The two subatomic parties continue to talk to each other, and whatever happens to one is identical to, or the opposite of, what happens to the other.

Entangled particles often get into a state of "coherence," where they lose their individuality and behave like one giant wave. Although the individual subatomic entities do maintain a certain individuality— much like the individual musicians in an orchestra—any attempt to separate them is now futile, for something occurring to one always affects the whole, and any action by one is conducted by the group. They are so inextricably connected that you cannot even distinguish one from the other. Frontier physicists have seen evidence that the subatomic particle emissions from living things, including human beings, are highly coherent, suggesting that our own subatomic makeup consists of particles that exist only as a collective, impossible to distinguish from each other.

Until his death in 2008, John Archibald Wheeler's sober countenance, not unlike a balding Richard Nixon, belied his almost recklessly speculative interior life. It was Wheeler, another protégé of Niels Bohr, who developed the theory of nuclear fission. He also came up with the idea that uranium-235 might make a feasible atomic bomb, well before that became the mission of the Manhattan Project in Los Alamos during World War II. It was also Wheeler who discovered that Einstein's space-time becomes a seething quantum foam, and who, with his talent for turning a phrase, coined terms like "black hole" and "wormhole." Wheeler acted as one of Einstein's chief cheerleaders in promoting the theory of general relativity and, working with Einstein, tried and failed to derive from it a unified field theory. Nevertheless Wheeler carried on believing the universe could be unified, presented as a line of mathematics, and even ultimately reduced entirely to information. Toward that end he produced the catchy phrase "it from bit." "Every particle," he said, "every field of force, even the space-time continuum itself is derived from 'yes or no' answers, binary choices, bits."[10]

Wheeler's most outrageous speculation attempted to understand the biggest mystery posed by quantum physics. The pioneers of quantum physics had demonstrated in their experiments that the only thing that seems to transform the potential something of a subatomic particle into something solid and measurable was the involvement of an observer. As soon as the scientists took a closer look at a subatomic particle by taking a measurement, the subatomic entity that existed as pure potential would "collapse" into one particular state.

The fact that a subatomic particle settles down to a set single state only when measured or observed suggested a staggering possibility to many scientists: the role of the scientist himself—or the role of living consciousness—somehow is the influence that turns the smallest elements of life into something real. This implies that the universe is a joint venture between the observing consciousness and the observed: it requires the observer to bring that which is observed into being.

Wheeler wanted to test this notion with a variation of the famous double-slit experiment in quantum physics, based on a classic experiment with light carried out by Thomas Young, a British physicist of the nineteenth century. In Young's experiment a beam of pure light is sent through a single hole, or slit, in a piece of cardboard, then passes through a second screen with two holes, before finally arriving at a third, blank screen. The light passing through the two holes forms a zebra pattern of alternating dark and light bands on the final blank screen. If light were simply a series of particles, two of the brightest patches would appear directly behind the two holes of the second screen, as a pattern of individual particles.

However, the brightest portion of the pattern is halfway between the two holes, caused by the combined amplitude of those waves that most interfere with each other. When two waves peak and trough at the same time, and bump into each other, the combined signal of the overlapped wave intensifies and the light gets brighter. If one peaks when the other troughs, the opposite occurs, and they cancel each other out, causing complete darkness. After observing this phenomenon Young was the first to realize that light beaming through the two holes spreads out in overlapping waves.

A modern variation of the experiment fires off single photons through the double slit using a gadget called an interferometer. These single photons also produce zebra patterns on the screen, demonstrating that even single units of light travel as a smeared-out wave with a large sphere of influence. Physicists have held up Young's experiment as proof that quantum entities like photons act wavelike and travel through both slits at once. Because you need at least two waves to create such interference patterns, the implication of the experiment is that the photon is somehow mysteriously able to travel through both slits at the same time and interfere with itself when it reunites.

But there is a catch to this experiment: when the experimental apparatus has a particle detector whose job is to locate which slit the photon went through, the outcome of the experiment is very different. Instead of being wave-like, the photon acts like a particle and is detected as definitely traveling through one of the two slits. Rather than creating an interference pattern, it creates a definite particle pattern on the screen.

So when the particle detector is turned on, it assumes the role of observer. Once "observed" by the detector, the photon acts like a solid particle rather than a smeared-out, uncongealed wave. It collapses to a single entity, goes through only one of the slits, and enables you to track its path.

In 1978, when Wheeler was pondering the meaning of this experiment—which seemed to place an emphasis on whether or not the photons were detected—he wondered whether timing was important, whether it mattered at which point the photon is observed or measured.

When scientists want to test a hypothesis, they sometimes first carry out a "thought" experiment, or, in a nod to the German roots of the practice, a *Gedankenexperiment*. They imagine an experiment in their heads and work it out in the pure language of mathematics. The experiment is thus essentially proven to work, or not, by virtue of math alone, not real life. To test the timing with his photon experiment, Wheeler devised a famous thought experiment called the Delayed Choice Experiment, in which a particle detector is delayed so that the photon's path is detected only after it has gone through the slits.

Imagine a photon that has already passed through the slits and is traveling toward the back walls. There are three possible routes for the photon: the left slit, the right slit, or both slits at the same time. At this stage we don't know which route it has taken.

What if, as Wheeler imagined, the apparatus could include a highly mobile detector screen, which can either be removed at this point or left in place? Once the screen is removed, two telescopes are revealed, each one trained on one of the slits. The telescopes would be able to see and record a little flash of light as the photon travels through one of the slits and so be able to detect the path of the photon through one or the other slit.

In this experiment the observer has "delayed his choice" of whether or not he wants to observe the path of the photon (via the telescopes) until after the photon has presumably made its decision to go through one slit, the other slit, or both.

According to Wheeler's calculations, the path of the photons entirely depends upon whether or not they are observed. If we remove the screen and the telescopes record the path of the photon—even after it has passed through the double slits—we get a distribution pattern consistent with what we'd get if particles were going through one or the other of the two slits, but not both. If we keep the screen up, the photons remain in a state of superposition and go through both slits.

The remarkable aspect of this experiment is that timing is irrelevant: even after the event has occurred and the photon has gone through one or both slits, the presence or absence of the screen—that is, the presence or absence of observation—determines its final outcome.[11]

Wheeler's experiment implies that observation, even after the fact, determines the final outcome. The observer entirely controls whether that which is observed comes into being at any point in time.

In the words of Wheeler's protégé, the physicist Richard Feynman, the role of the observer is central to quantum physics; it is the "mystery which cannot go away." Wheeler's experiment itself remained a mystery until 2007, when Jean-François Roch and his colleagues at the Ecole Normale Supérieure de Cachan in France successfully found a means of carrying out the "delayed choice" experiment and proving the ideas that Wheeler had imagined thirty years before.[12]

As Wheeler noted in 2006, two years before he died, "We are participators in bringing into being not only the near and here but the far away and long ago."[13] In his fertile imagination, he even imagined the entire universe as one giant wave in need of observation to bring it into being.

From this evidence we have to ask ourselves a fundamental question: If quantum entities, which are impossible to separate from each other, drive all our basic life processes, does anything exist as an actual *something* on its own?

Matter in the subatomic world cannot be understood in isolation but only within a complex web of relationships, forever indivisible. Life exists because of a fundamental duality, a multiplicity of influence and being, a cooperative partnership. At its most elemental, physical matter not only isn't an *anything* yet, but it remains something indeterminate until our consciousness becomes involved with it. The moment we look at an electron or take a measurement, we help to determine its final state.

The most irreducible relationship of all may be that between matter and the consciousness that observes it; what finally makes anything real is the alchemical Bond between observer and observed. There is no "us" and "them," only a constantly transforming "we." With every breath we take, we are cocreating our world.

We cannot discover the most fundamental parts of our universe, no matter how hard we look, because they exist only in relationship to the other parts. Quantum physicists fruitlessly continue to hunt for the thing, even though, in the very act of searching, they alter it. Life is established not within a thing but in the Bond, the space between two things: between subatomic particles, between particles and the background Field, and between mind, or consciousness, and matter. In fact, as biologists have discovered, that is the very way we ourselves get made. You and I are creations entirely of our interactions with the universe.

16

CHAPTER 2

THE MASTER PROGRAMMER

To Randy Jirtle, the human genome was simply a faulty computer. Jirtle, a professor of oncology at Duke University, had begun his long academic career working on computers, performing feats of complex mathematics as part of his undergraduate degree in nuclear engineering. After graduation he jettisoned his initial plans to work on nuclear energy and reactors, eventually specializing in the biology of ionizing radiation—the biological effects, in a sense, of nuclear energy gone wrong. But even with living systems it seemed like he was back to working with hardware and software. It was clear to him that the gene was the hardware. But why, he often wondered, did the hardware so often go wrong? And where exactly was the software, the master program controlling it all?

The flaw in the software had to do with *imprinting,* when one entire chunk of DNA on every chromosome pair gets silenced for life. Imprinting has baffled many geneticists, for it seems to violate the neat laws of inheritance laid down by Gregor Mendel, a nineteenth-century monk who developed the foundation for modern-day genetics. Although an organism inherits two sets of genes from each parent, according to Mendel, the tougher and fitter gene is *dominant,* helping to shape the organism's appearance, while the *recessive* other half, the ninety-pound weakling, stays silent, overwhelmed by the dominant gene. Nature's rationale for receiving two genomic copies from both mother and father appeared to be an extra safeguard against mutation, but by silencing one or the other, nature seemed to be relinquishing this advantage.

Jirtle contended that the determination of whether or not a gene turned on had more to do with which parent it had come from than

17

any innate tendency to be dominant. At the level of the gene, life is a constant war between the sexes. As his own work revealed, the silenced gene appeared to be inherently weak, a susceptible target for cancer or other disease. It was as though DNA had been constructed with a circuit board deliberately programmed to implode.

For an entire decade Jirtle had been puzzling over this issue and how it related to systems in the body that suddenly developed cancer. His laboratory was noted for its investigation of insulin-like growth factor-2 (IGF-2), a substance in the body that stops apoptosis, or cell death, paving the way for cancer: cell growth run amuck. He'd exhaustively studied a tumor suppressor called IGF-2R in mice, which scavenged for IGF-2 and managed to mop it up, ultimately suppressing liver cancer. Jirtle's team eventually discovered that a switch *outside* the gene was responsible for turning on IGF-2R activity. But what exactly were the biological conditions that flipped the switch? If he could find the on and off switches for certain diseases, he might be able to modify the imprinting process and ultimately turn off the computer's self-destruct mechanism.

Jirtle had his chance to explore the question further in 2000, when he received a letter from Rob Waterland requesting sponsorship for his postdoctorate project. Waterland, who'd just received his Ph.D. in human nutrition from Cornell University, wished to explore whether dietary changes represented one of the switch's main controls, and he'd been impressed by Jirtle's willingness to look beyond the gene as the master control. Waterland had been particularly fascinated by the scientific evidence showing that the children of women who were deprived of certain nutrients while pregnant were more highly susceptible to disease and neurological disorders. Epidemiological research had revealed that populations exposed to famine prenatally have lower birth weights and higher than normal rates of degenerative diseases, including diabetes, coronary heart disease, and cancer.

Waterland was also intrigued by evidence that the effects of famine carried on through generations. People who had been starved in the womb produced smaller than normal children, who in turn had smaller grandchildren, even if the children and grandchildren received adequate nutrition throughout their lives. Adverse environmental conditions appeared to affect at least two generations down the line.[1]

Suppose you attempted this in reverse, he wondered. If you *improved* the diet of a pregnant mother, could you turn off the expression of genes programmed for disease and reverse genetic destiny?

Jirtle agreed to work with Waterland, with the proviso that they choose their model species carefully. When scientists want to study a genetic problem, they need to examine an easily identifiable aberration under exaggerated conditions, so they look for a genetic quirk of fate. Jirtle and Waterland both independently decided that the perfect quirk for their purposes was the agouti viable yellow (A^{vy}) gene. Agouti mice, as they're known, possess a genetic defect in their agouti gene, which encodes the signaling molecule for hair color, ordering the hair follicles to produce a yellow coat rather than the customary brown. Mice with the defective agouti gene are also destined to be couch potatoes. Besides being golden colored, these mice are often hugely obese, with a tendency to develop diabetes and cancer.

Waterland had been inspired by a study carried out by the National Center for Toxicological Research (NCTR) showing that supplementing the diets of female agouti mice with a variety of B vitamins could override the genetic defects and produce a greater number of normal offspring. The study had speculated that a mechanism "above the gene" had been responsible, but the researchers had not identified exactly how it worked or what in the mother's DNA got modified to produce this happy result.[2] Waterland planned to use their model and their protocol, but then take apart the offspring mice's DNA to see what it might reveal.

After receiving a two-year fellowship from the Dannon Institute, Waterland and Jirtle got hold of a few pairs of breeding agouti mice from the NCTR group who had done the original study, and within six months had ten litters each of test mice and controls. Half of the female agouti mice were fed extra B vitamins before they became pregnant, during pregnancy, and also during lactation, while the other half received only normal mouse chow.

Isolating the genetic code of a single animal is a painstaking process that can take an entire week. Waterland started by taking a tiny fragment of tissue from the tail of each of his test animals, modified it through one process, mixed it into a toxic soup of chemicals to separate out the genetic code, then shocked it with 40°C changes of temperature. The alternating hot and cold set off a chain reaction that quickly replicated the DNA, a bit like a photocopier firing off endless copies. After a few more chemical processes, the genetic blueprint was finally visible enough to photograph. Once he had carried out this laborious sequence with all his test mice, Waterland discovered an obvious difference in an aspect of the code of the mice whose mothers had been supplemented.

Genetic code for proteins comes in four versions, notated by science with uncharacteristic simplicity as A, C, G, and T (representing the nucleotide bases of adenine, cytosine, guanine, and thymine). In a huge portion of the offspring of the mice who'd been given the B vitamins, C had transformed into T. The B vitamin supplements, in a sense, had turned on a different gene.

The changes in gene expression were also physically obvious. A larger percentage of offspring of the mothers that had received the enriched diet were a normal brown color and also less susceptible to adult degenerative diseases like diabetes and cancer. Unlike their mothers, the next generation of mice lived out a normal life span.[3] The dietary supplements had dramatically overridden the offspring's genetic destiny by turning off the agouti gene's expression. This was the first evidence of a clear causal pathway between a mother animal's environment and permanent alterations in an aspect of her offspring's genetic code. As Jirtle giddily announced with the publication of their study, "This is where environment interfaces with genomics."[4] With their little batch of mice Jirtle and Waterland had proven that a few simple environmental modifications early in the life of a living thing could take charge of its genetic destiny.

"We're either going to be famous or we're going down in flames," Jirtle had joked to Waterland when the journal *Molecular and Cellular Biology* featured their study on the cover of its August 2003 issue, accompanied by a photo of an assortment of yellow and slim, brown offspring crawling over a mountain range of fresh broccoli and kale. Jirtle realized what they were about to unleash upon the scientific community: not only had they overturned a century's worth of scientific belief about genetics and the backbone of modern biology, they had also redeemed Jean Baptiste Lamarck, a scientific maverick so disparaged that for more than a century his name had been a synonym for the preposterous.

*　　*　　*

More than fifty years before Darwin developed his views on natural selection and wrote *On the Origin of Species,* Lamarck, a French botanist and zoologist, had spent ten years studying French flora and invertebrates (a term he coined), eventually producing *Recherche sur L'Organisation des Corps Vivants* in 1802 (*Research on the Makeup of Living Bodies*) and later his masterpiece, the two-volume *La Philosophie Zoologique* (*Zoological Philosophy*), the first book to set out a coherent, well-developed theory of evolution.[5] Lamarck described life as an evolving chain of being, and indeed Darwin enthusiastically studied Lamarck's theories as a student at Oxford. Yet Lamarck also believed in the inheritance of "acquired characteristics," that the environment is responsible for changes in animals, and that these changes could be inherited by the next generation.

Darwin depicted the evolution of species essentially as capricious happenstance. The whole of his theory of natural selection rests on three basic assumptions: that all organisms descend from a common ancestor, that new traits in species evolve by random mutation, and that these traits persist only if they help the species to survive. He believed that mutations occurred within individual organisms essentially as some sort of copying error passed down to offspring. Any permanent changes resulting from these mutations would endure within a species only if this genetic mistake provided members of the species with a survival advantage. Nevertheless any major changes from one species to another occurred in infinitely small baby steps over vast stretches of time. From this perspective, as Richard Dawkins memorably put it, nature is a "blind watchmaker" and evolution a cold-blooded process of winnowing out winners from losers.

Lamarck, on the other hand, viewed evolution as a cooperative venture between an organism and its environment. He believed in the "inheritance of acquired characteristics," that an organism acquiring certain characteristics during its lifetime *in response to environmental challenges* would pass these traits on to its offspring. He concluded that an organism responded to the *need to evolve* and that this need would produce favorable adaptations. Like Darwin, Lamarck theorized that "transmutation of species," as he called the evolution of one species into another, occurred slowly, over many eons. Unlike modern evolutionary synthesis, with its emphasis on a genetic roll of the dice, the Lamarckian worldview depicts the natural world as a dynamic, symbiotic partnership, and evolutionary change as a joint solution, restoring balance and harmony when an organism is out of alignment with its environment.

Although Lamarck's views were initially greeted with enthusiasm, eventually they were roundly rejected, and he died a pauper, his body buried in a lime-pit grave. "Lamarckian" entered the scientific lexicon as a disparaging term for any idea suggesting that environmental factors could alter genetic coding or that an organism of any variety somehow had the intelligence to affect the body's central blueprint.

It took more than a century for Lamarck's work to be reexamined, when a British embryologist and Cambridge University lecturer named Conrad Waddington, while attempting to work out how the neural tubes of amphibians get constructed, grew convinced that the answer lay in the fledgling discipline of genetics. This was a heretical thought in the 1930s, because genes were not fully understood and certainly not considered a hereditary blueprint for anything other than basic physical characteristics.

To pursue his rogue theory, Waddington traveled to the American West Coast. At the California School of Technology the great American geneticist Thomas Hunt Morgan had created a laboratory entirely devoted to *Drosophila melanogaster,* the common fruit fly, which would prove to be, under Morgan's work, the model organism of study for generations to come. Once settled at the California lab Waddington began to painstakingly analyze how tiny mutations affect the development of the *Drosophila's* hind wing by exposing the embryonic flies to ether.

Although Waddington originally supposed his inquiry would prove the primacy of genetic coding, eventually he discovered that abnormal conditions at early stages of a fruit fly's life would cause it to develop a strange set of hind wings. After numerous generations of flies had been exposed to ether Waddington discovered something else remarkable. The altered hind-wing development, a purely environmentally induced change, continued to get reproduced in eight generations of offspring, even when they were not exposed to the volatile liquid.[6]

Waddington went on to coin the term "epigenetic landscape" to suggest that the environment helps to complete genetic expression. He used the term "epigenetic," which means "above the gene," because this influence appeared to occur outside the gene itself. He also worked out his theory of *genetic assimilation*, which proposed that an animal's response to its environment, particularly to stressors, not only affected its development, but also became part of its inheritance; the resulting acquired characteristics created changes that were assimilated into the species, producing major changes for many succeeding generations.[7] Waddington was the first to demonstrate that Lamarck, and not Darwin, had been right in one regard: the development of a living thing appeared to be dependent upon the nature of its connection with things outside itself, and it was this *connection,* and not simply the genetic code, that got passed down to subsequent generations.

The most fundamental proof we have of our own distinction lies in the fact that our body is unique from all others. Our concept of *self* also stems from the belief that our body is created through an exclusive and entirely self-sufficient process within its own boundaries; personality, physical characteristics, indeed the sum total of what defines us is crafted from the unique blueprint of DNA contained inside. Although we allow for the effect of emotional stress on our personal psychic development and of diet on certain aspects of our health, we assume that the raw clay of ourselves takes permanent shape and then sets and hardens largely from a process that moves outward from the gene through our cells to our organs. The course of our lives, good or bad, is not thought to alter either our own genetic blueprint or the one we pass on to our children, other than through random mutations occurring over hundreds of generations.

Ever since the natural philosopher (and Newton's archenemy) Robert Hooke peered through a primitive microscope in 1660 and discovered what appeared to be self-contained units in pieces of cork, biologists have considered the cell to be the engine room of the human body. Indeed the word "cell" actually means "small room" in Latin, for Hooke decided that his cork cells resembled a monastic cell, with its mitochondria—the actual engine—and the nucleus, the headquarters of operation. Nevertheless it required three centuries of scientists staring inside this tiny subsection of a cell to discover what came to be regarded as the body's central engineer.

After Watson and Crick unraveled DNA, the genetic coding within the cellular nucleus, DNA was looked upon as the architect, builder, and overlord of the body, drafting a lifelong blueprint and then using it to spearhead and oversee all the body's dynamic activities. This was all supposedly managed through a straightforward, mechanistic process of selectively turning on and off certain genes, the steps on the spiral ladder of the double helix. These nucleotides, or genetic instructions, make copies of themselves as messenger ribonucleic acid (RNA) molecules, which choose from an alphabet of amino acids the genetic "words" that create one of the body's approximately 150,000 specific proteins to carry out its myriad functions.

From the outset Watson and Crick laid down a set of rules they termed the "central dogma" of microbiology, as if to underscore its acceptance as an article of faith. The first commandment was that cellular informational commands flowed in a single direction, from DNA to messenger RNA to the selected combination of amino acids to the assemblage of proteins. Any proposals with the faintest suggestion that this process was reversible—that outside agents could affect the proteins that make up living organisms, and that these proteins could affect RNA and ultimately DNA—were discredited as Lamarckian fantasy.[8] Nevertheless, the central dogma fell short when it attempted to explain exactly how this long chain of genetic instructions "knows" when to orchestrate a particular process and what exactly provides the signal. Until recently scientists maintained that gene activity is a hermetic process, occurring independently of the environment.

As Jirtle and Waterland's study decisively demonstrated, genes, far from being the central controller, exist much as subatomic particles do, purely as a potential,[9] to be activated or not by signals outside our body. Research now suggests that information actually flows from the other direction: *from outside in*. An environmental signal of some sort alerts the body that a particular protein product is needed, and it is the outside environmental signal that activates particular genetic expression. The intricate array of environmental influences to which we are exposed throughout our lives actually determines the final expression of every gene in our body. Genes get turned on, turned off, or modified by our life circumstances and environment: what we eat, who we surround ourselves with, and how we lead our lives.

*   *   *

Imagine a giant manufacturing plant with a central office and large numbers of energy centers used to power the rest of the plant, so large and sophisticated that they are host to thousands of simultaneous chemical and electrical processes. Then imagine that there are 40 trillion of these extraordinary manufacturing plants sitting cheek by jowl, trading resources back and forth, and you begin to comprehend the dynamic life of every one of the 100 trillion cells in your body.

Each cell is a body unto itself, capable, in the space of ten micrometers, of carrying out all the varied activities—respiration, consumption, replication, excretion—that your body engages in as a totality. Nevertheless, no matter how adept at multitasking, acutely observant of change, and rapidly adaptable, not one single cell in your body is capable of any function without receiving a signal from outside itself. In fact, as scientists are now beginning to understand, the switch that turns genes on or off lies outside your body.

The cytoplasm, the blob of jelly that makes up every cell in your body, is encased in a semipermeable cell membrane, a triple layer of fat-like molecules containing a variety of protein molecules that act as little revolving doors for other molecules to enter or exit the cell. Whether or not a molecule gets through the cell membrane depends entirely on these gatekeeper proteins, which are called *receptors* because they function like antennae, picking external signals from other molecules and in turn signaling to *effector* proteins to modify the cell's behavior.

The membrane contains hundreds of thousands of these protein-receptor switches, which possess the ability to regulate a cell's function by switching a certain gene on or off. But, as Jirtle discovered, what prompts the turn of the switch is an environmental signal—from the air, water, and food we consume, the toxins we're exposed to, or even the people we surround ourselves with. This in turn affects the methylation of the DNA double helix, which is exquisitely sensitive to the environment, particularly in the early stages of life. During this process the methyl group, a quartet of atoms, attaches to a specific gene and sends it a message to silence it, reduce its expression, or in some way alter its function.

Science used to regard this configuration, often termed the "epigenome," as merely responsible for cell differentiation, ensuring that even though every cell carries identical DNA, certain cells turn into a nose and others into an arm. But as the agouti mice study demonstrated, the epigenome also has the potential to function as an interface between outside and inside the body, as the gene's interpreter of environmental signals. The B vitamins in Jirtle and Waterland's study acted as methyl donors, causing methyl groups to attach more frequently to the agouti gene of baby mice in utero, turning off its expression as a trigger of rogue golden coloring and eventual disease. This signaling occurs not within but *on top of* the gene and does not alter the genetic sequencing or interfere in any way with the letters of the four-unit genetic code. This means that an important controller of a gene and whether or not it is activated is the host of influences outside the body.

Stripped down to its basics, one cell of any person or animal is indistinguishable from any other. Take the membrane off one of my 40 trillion cells and you can successfully implant it into your body.[10] A cell has no individuality unless it interacts with its environment. Outside influences will determine a cell's expression and how it reacts within its world, whether it will conform or be an outlaw to its fellows. As the biologist Bruce Lipton notes in his groundbreaking book, *The Biology of Belief,* the real brain of a cell is its membrane.[11]

Rather than the nucleus, external influences filtering through the cellular membrane control the cell, and consequently the behavior and health of the entire organism. Epigenetic changes and the ultimate expression or silencing of a gene occur as a result of environmental stressors. Diet, the quality of air and water, the emotional climate within your family, the state of your relationships, your sense of fulfillment in life—the sum total of how you live your life and also how your ancestors lived theirs—have the most effect on the expression of your genes. Every factor in our lives conspires to create the physical person we become.

Discoveries in epigenetics represent the ultimate apostasy to Watson and Crick's central dogma that genes determine how we respond to our environment. As Jirtle and Waterland's experiment showed, the reverse occurs: Our environment determines how our genes will respond. The environment *outside* our bodies determines the environment *within.* We are constructed by a delicate interplay of substances, inside and outside of our physical borders. A Bond exists between the blueprint within our cells and everything we connect with in our world throughout our lives. Like subatomic particles, our physical body is not a discrete entity, but the end product of a relationship.

Jirtle and Waterland's mouse study set off a riot of research. Five years after its publication the number of papers published on epigenetics grew fortyfold, particularly those concerning inheritable diseases. At the forefront of this research was a team at McGill University in Montreal led by Moshe Szyf, an Israeli-born professor of pharmacology and therapeutics. Szyf's lab owns five patents on DNA products and one patent pending, all for DNA formulations that he hopes will change the course of medical history. He believes that he may find the cure for cancer within the human epigenome by manipulating the methylation process so that the on-switch for cancerous genes gets turned off permanently.

Szyf has discovered that a major hallmark of cancer is an aberration in methylation patterns, so that genes needed for rapid cell growth, invasion, and metastasis aren't kept in check.[12] Although other researchers think cancer arises from too much methylation around a gene, Szyf believes the problem has to do with both too much and too little; too much methylation in breast cancer, for instance, silences genes necessary to regulate cell growth, and too little tends to activate genes involved in rapid metastasis. Szyf's patents cover products that will regulate the methylation process in individuals with cancer, a process he's been able to demonstrate in human cancer cell lines in the lab.[13]

Szyf's work defies current thinking about epigenetics. Many scientists exploring this new field had first assumed that epigenetic changes operated a bit like the butterfly effect in chaos theory, with its sensitive dependence on initial conditions; small changes occurring early on in your environment when you are a baby will produce large changes in genetic expression, but then remain constant throughout life. Szyf's work in the laboratory decisively demonstrates otherwise.

In a series of studies on animals he showed that numerous kinds of stress responses deliberately programmed into a variety of animals by one set of conditions early in life could be deprogrammed out of the organism by changing the conditions later in life. In one study Szyf was able to reverse abnormalities in baby rats caused by unhealthy mothering by handing the rat pups to foster mothers who treated them normally. Epigenetic conditions now appear to be fluid—wholly reversible in adulthood.[14]

A disease like breast cancer may also have its genesis in the Bond between our inner and outer worlds, and not exclusively within our genes. Of all forms of cancer, a family history of breast cancer is usually assumed to be one of the most clearly marked genetic indications that a woman is likely to develop the disease. Recently some doctors have counseled healthy women with a certain gene to undergo a single or double mastectomy to prevent the development of breast cancer.

Several epidemiologists at the University of Rochester Medical Center in Rochester, New York, questioned this practice after examining data from one of the most controversial studies of women in American history, the Women's Health Initiative (WHI), sponsored by the National Institutes of Health. Among the largest studies to follow women for several decades, the WHI was expected to confirm the safety and benefits of hormone replacement therapy (HRT), as well as other treatments and practices. Five years into the study of hormone replacement the Data and Safety Monitoring Board of the WHI shocked the world by calling an abrupt halt to the use of HRT when it became apparent that the 16,000 participants who were taking hormones had an increased risk of developing breast cancer, ovarian cancer, stroke, and heart disease.[15]

For the Rochester scientists, the WHI represented a goldmine of data for comparing hereditary and environmentally induced cancer. When they combed through the details of the women taking part in the study who had developed breast cancer, they naturally assumed that they would find a higher incidence of cancer among those who had a family history of the disease. However, the evidence showed a similar incidence of cancer among those taking HRT, whether or not they had breast cancer in their genetic history. The particulars of a woman's genetic makeup or a family history of cancer appeared to have nothing to do with it.[16] In this case the environmental stressor—artificial hormones taken regularly—was the major trigger.

Another kind of Bond with the same capacity to affect genetic expression is the quality of our social relationships. Moshe Szyf examined and compared the brains of suicide victims deliberately chosen for having had an abusive or neglectful childhood with those of patients who had died from ordinary causes. Although the genetic sequence was identical in both sets of brains, fascinating differences appeared in the epigenetic markings of the genes within the brains of those who had committed suicide.[17] Szyf could not categorically conclude that the abuse in childhood definitely caused both the epigenetic markers and ultimately the suicidal depression, but the circumstantial evidence was compelling.[18] His findings were echoed in recent work at the Centre for Addiction and Mental Health in Toronto on patients with schizophrenia and bipolar disorder. These patients were found to have alterations in the outer casing of neural DNA, again strongly suggesting an environmental cause of the mental illness, and not inherited genetic history.[19]

The Bond we make between our inner and outer worlds is so powerful that it can also have a positive effect, compensating for a bad genetic hand. Li-Huei Tsai, a neuroscientist and director of the Picower Institute for Learning and Memory at the Massachusetts Institute of Technology, investigates the neural architecture that is involved in memory in hopes that her work will lead to a way to prevent brain degeneration. In 1997 Tsai began to explore whether epigenetic changes could improve memory, particularly in a mammal with genetically induced brain damage.

Tsai and her team selectively bred a group of mice with a certain genetic expression called p25, which causes neural degeneration and eventually a condition akin to Alzheimer's disease. Animals with this genetic configuration have profoundly impaired learning and memory. In very short order, after brain atrophy and progressive loss of neurons, these mice quickly become demented. Impressed by studies showing that an enriched environment can improve learning capability, Tsai decided to test whether this would apply to an animal that had already suffered brain degeneration.

In the first test every time the mice went into a specific chamber of their cage to carry out a particular task they received a mild electric shock. In the second test the mice had to locate a platform they'd seen before, but that now lay submerged in murky water. As the mice could no longer see the platform, they had to rely entirely on their memory of its whereabouts.

Ordinarily the fear conditioning of the electric-shock test produces a long-term memory of the event; once a child burns his hand on the stove, he tends to remember in future to steer clear of a gas flame. Tsai's group of mice failed both tests; their brains had atrophied to the point where they could not learn from an unpleasant experience or retrieve from memory where an object might be located.

For the next series of tests Tsai placed her population of mice in an action-packed environment, containing an exercise treadmill, toys of different shapes and textures that were changed every day, and a new group of mice. After the experimental mice had spent time in their new environment, the researchers ran them through both tasks again. This time the mice demonstrated an ability to remember the electric shock chamber and the submerged platform. When Tsai and her colleagues later studied the brains of the test animals, they discovered that the environmental stimulation had modified epigenetic chemical tags and methyl groups called histone acetylation, which ultimately turned off the expression of their genetically modified Alzheimer gene.[20]

The mice experiments show that, even with something as complex as genetically induced memory impairment, genes are not destiny. In this case the Bond between the animals and their stimulating environment overrode the animals' genetic handicaps.

Larry Feig and his research team at the Sackler School of Biomedical Sciences at Tufts University carried out a similar experiment with a group of mice with a silenced Ras-GRF-2 gene, which also affects memory and learning. But this time the team focused on the offspring that were bred during the two-month window when the mother mice lived in their "theme park" and enjoyed normal learning ability. According to genetic theory, the offspring should also have inherited the silenced Ras-GRF-2 gene, particularly as they were raised in a normal laboratory environment rather than the stimulating theme park.

Astonishingly the offspring of these mice showed evidence of normal memory and learning ability—even though they themselves had had no additional stimulation and should have inherited the turned-off gene. The environmental conditions of their mother overrode their genetic destiny.[21]

Like the agouti mouse study, Feig's work suggests that the Bond between a living thing and its world, and not simply the genetic code, is ultimately what gets passed on to subsequent generations. A mother's positive diet and positive environment represent a far more potent inheritance for her children than "good genes."

The reverse happens as well: we can also inherit the effects of our parents' negative environment. Lars Olov Bygren, a specialist in preventive health at the Karolinska Institute in Stockholm, examined nearly one hundred individuals born in the Overkalix parish of Norrbotten, the tiny northernmost county of Sweden. Bygren knew that the isolated, sparsely populated area had suffered unpredictable harvests in the nineteenth century; in 1800 and 1821, for instance, the crops had failed and its inhabitants had starved, but in 1802 and 1822 the harvest had been so abundant that the Norrbotten people who'd gone hungry a year or two earlier gluttonously feasted on food for months. Bygren wondered about the long-term effects of these boom-and-bust food cycles on the offspring of the Norrbotten inhabitants. After scouring historical and agricultural records he was able to determine that those men and women who had turned gluttonous for a single winter produced children and grandchildren who had lived shorter lives.[22]

Bygren tested these conclusions by teaming up with the British geneticist Marcus Pembrey of University College, London, to carry out a large epidemiological study in the county of Avon in Great Britain. Within the total study population of more than 14,000 the researchers found 166 men who had started smoking before age eleven, which is just before the time boys enter puberty and begin to produce sperm. When Pembrey and Bygren examined the records of the children in the study, they found that at age nine the sons of the early smokers had significantly higher body-mass indices than the sons of fathers who were not early smokers.[23] The Avon group also discovered that mothers who pile on too many pounds during pregnancy predispose their offspring to future heart risks.[24] Just one environmental stressor at this key point in life can have a major impact on the future health of offspring.

One of the greatest of environmental switches may be the quality of our Bond with a social group. Psychologists at Northwestern University have examined the effect of social grouping on genetically inherited predisposition to depression. True clinical depression, like most illness these days, is considered largely the fault of a bad toss of the dice. The entire edifice of standard treatment for depression rests upon the theory that depression results from a chemical imbalance within the brain, considered to be largely hereditary.

The main genetic component for recurrent depression has to do with the serotonin transporter gene, which comes in two distinct varieties, the short allele and the long allele. The short allele, as its name suggests, represents the short straw. This variation carries the major "on" switch for depression; anyone with this gene who experiences a series of major life stresses is considered a prime candidate for major depression.

The Northwestern team is part of a new field called cultural neuroscience, which examines mental health across nations and individual social groups. One of the greatest distinctions in any culture is how someone thinks of himself in relation to other members of his society, and whether he views himself as mainly "individualistic" (a lone wolf) or "collectivistic" (a small cog in a greater whole).[25] The team examined health in relation to the cultural values of a cross-section of countries on most of the world's continents, focusing on the degree to which each population places greater emphasis on the individual or the group.

In a nutshell, they found that Westerners define themselves by their individuality and Easterners define themselves by the extent of their acceptance within a group. "People from highly individualistic cultures like the United States and Western Europe are more likely to value uniqueness over harmony, expression over agreement, and to define themselves as unique or different from the group," said Joan Chiao, the team leader.

In collectivistic societies, such as those of East Asia, on the other hand, higher value is placed on social harmony than on individuality. The culture encourages behaviors and practices that endorse interdependence and group cohesion.

Chiao's team made an unexpected discovery: the more tightly knit the population, the higher percentage of the people who carried the gene for depression. East Asia in particular has a hugely disproportionate number of carriers of the short allele; at least 80 percent of the population are genetically susceptible to depression. According to the current genetic theory of depression, correspondingly high levels of depression should exist among these populations. Instead Chiao found the opposite: among these highly susceptible populations, the actual prevalence of depression was significantly lower than in Western Europe or America.

The expectation of social support in these highly collectivist cultures seemed to buffer people from any environmental stressors that should have triggered depression.[26] Even genetically inherited depression could be controlled by a social switch.

In the mid-1980s John Cairns, a British-born geneticist at Harvard's School of Public Health, carried out an experiment that would set off one of the largest arguments in modern biology. The plan of the experiment was simple enough: to place some bacteria in a tight spot. Cairns selected bacteria with a genetic defect rendering them unable to digest lactose, the sugar present in milk, then introduced them into a batch of Petri dishes containing cultures whose only food source was lactose. Without any digestible food, the bacteria faced death by slow starvation.

According to orthodox science and the neo-Darwinist view of natural selection, the bacteria would not be able to colonize; without a food source to drive metabolic processes, they could not carry out normal reproduction. Nevertheless in every Petri dish Cairns found a goodly number of thriving colonies.

When Cairns tested for genetic changes in his colonies, he found that a single type of gene had changed: those preventing lactose metabolism. Identical changes in just those genes had occurred within every new colony in every Petri dish. Cairns had confirmed that none of the original bacteria contained a lactose-digesting mutation prior to the experiment. Through some unknown mechanism the bacteria had activated eleventh-hour mutations in direct response to an extreme environmental crisis, and these mutations had saved their lives. The bacteria had defied the central dogma: they had evolved purposefully, not randomly, in order to restore balance and harmony with their environment. Somehow the extreme environmental conditions had caused changes in genes, enabling the bacteria to digest the only food available to them.

In 1988 Cairns published his findings in the prestigious journal *Nature* under the droll title "The Origin of Mutants," a flippant nod to Darwin.[27] He proposed that cells within organisms have the ability to orchestrate their own "directed mutation," rapidly adapting to a changing environment. Although Cairns, the discoverer of the structure and replication of the *E. coli* genome, had great standing among his peers, the Lamarckian assertion that the environment changed genes sparked a decade-long protest in the medical literature. The American journal *Science* dismissed his work as tantamount to "heresy."[28]

As other researchers looked closer they discovered that, under environmental stress, a special enzyme in a bacterium cell gets activated, initiating a fevered copying process of cell DNA with a deliberate array of random mistakes, a mechanism now referred to as *somatic hypermutation*. If any one of these mutated genes is able to assemble a protein with the key to overcoming the environmental problem, the unthinkable occurs: the bacterium jettisons the original problem gene from its DNA and replaces it with the new gene.[29] This is the likely process by which bacteria continuously manage to outwit antibiotics. Although Darwin described mutation as a random accident during the reproductive process, Cairns and other scientists after him have shown that the environment is constantly changing an organism, not simply through epigenetics but directly *by changing genes*.

Scientists have refined Cairns's early ideas after discovering that information between genes, the rest of the body, and the environment flows in a dynamic and interactive manner. "The traditional mechanistic view held that the structure of biological molecules determines the actions of cells in some kind of linear fashion," says James Shapiro, a professor in the Department of Biochemistry and Molecular Biology at the University of Chicago. Genes change, says Shapiro, not from accidents, but through "natural genetic engineering," or "adaptive evolution," as the process is now called, a constant dynamic process of adaptation between an organism and its environment.[30] "Today we know that biological molecules change their structures as they interact with other molecules and these structural changes contain information about the external environment and conditions within the cells."[31]

All of the recent research on adaptive mutation and epigenetics casts a long shadow on the idea that illness is simply a case of having "good" or "bad" genes. Not only are the on-off switches for genetic expression controlled by environmental triggers, but disease of many varieties—cancer, inherited defects, dementia, suicide, schizophrenia, depression and other mental illnesses—all appear to be set off by influences outside our bodies. Diet, a strong social network and community ties, purposeful work, mental stimulation, and an environment free of toxins and pollution may be far more important than the genes you are born with in determining the person you become and how healthy you are.

The implications of Jirtle and Waterland's mouse study extend beyond health and illness. With one small study they effectively demolished the central edifice of molecular biology, the mechanistic assumptions about the central workings of living things, including the primacy of genetic information. Genes can no longer be considered the only driving force of nature if an outside influence can completely derail the programming. Studies such as Bygren's and Pembrey's also showed that new traits can show up in a single generation—for better or worse—depending upon the nature of the relationship between the parent and his environment. The Bond between a living thing and its world, the relationships we have with each other and our environment, are the greater hereditary force. This information turns the orthodox concept of evolution on its head; rather than random accident, it is a cooperative process, a finely tuned and constant striving for harmony between a living thing and its world.

Epigenetics and adaptive evolution also display something remarkable about how we take physical form. The relationship between a living thing and its environment is a two-way, ongoing conversation. Although much of that conversation is set down early in our lives, it is dynamic, fluid, even reversible—a relationship for life. We are a balance of internal and external influences, early and late programming, constantly transformed by the influence of every moment.

These new discoveries beg the unsettling question: Where exactly do "you" end and the rest of the universe begin? If you internalize and change with every interaction you have with the universe—every bit of food you eat, every person you meet, every place you've visited—what exactly does it mean to be you? How can you be considered autonomous?

What we regard as *self* is only a physical manifestation of our experience, the summation of our Bond with the universe. Our interaction with our world is a conversation, not a monologue; just as the observer changes what he observes, that which is observed changes the observer. And as we now realize, these influences are not limited to those of our immediate environment or even to the earth itself but extend to the farthest reaches of the cosmos.

CHAPTER 3

THE *ZEITGEBER*

In 1922 Alexander Chizhevsky, a young Belarusian scientist, unveiled to the world a preposterous theory: that all the great upheavals in the history of man, such as social unrest, war, and revolution, were caused by the activity of the sun. These extraordinary claims, contained in his first book, *Physical Factors of Historical Process,* were greeted with near universal derision, and for a time the brooding, pallid twenty-five-year-old, a descendant of a court tenor and a member of heredity nobility who was already tainted by his aristocratic lineage in the eyes of his countrymen, became the laughingstock of the newly ensconced Bolshevik Party, which disparagingly nicknamed him "the Sun Worshipper." After all, he was essentially suggesting that all the tumultuous events leading to Russia's liberation from corrupt tsarist rule had nothing to do with ideology or the motivation of the country's workers and everything to do with sunspots.

As a consequence Chizhevsky fell out of favor for years among the scientific establishment, despite the patronage of the Nobel Prize winner, author, and political activist Maxim Gorky. Nevertheless Chizhevsky, a scientific da Vinci of sorts, stubbornly carried on with his research, attempting to make connections between biology, physics, geology, and space weather that remained invisible to most of his peers. He painstakingly examined records of all battles, upheavals, riots, revolutions, and wars, comparing them with sunspot activity for nearly two thousand years in seventy-one countries, including his own. The theory checked out: more than three-quarters of all instances of human unrest, including the Russian Revolution of 1917, had occurred during a solar maximum, the period of the maximum number of sunspots in any solar cycle. The only area that remained questionable was the

37

mechanism of this cosmic connection, but Chizhevsky had a theory: our dependence upon the cosmic pulse of the sun might be mediated by ions, or excess charge, in the air.

Chizhevsky may have been influenced by a French physicist named Jean-Jacques d'Ortous de Mairan, who'd discovered that one of his plants folded up its leaves and "went to sleep" at the same time every night, even when placed in total darkness. The mechanism for this strange activity was right under de Mairan's nose, but at the time he didn't see it. Although he'd written a book about the aurora borealis, he dismissed the idea that solar activity and magnetism could be responsible for the regular timetable of his plant. Two centuries later Chizhevsky immediately understood the connection.

Although the Soviet government eventually provided Chizhevsky with his own laboratory, largely because of his contribution to the understanding of air ionization, one person who remained singularly unimpressed with Chizhevsky's theories was Josef Stalin, who in 1942 demanded that the scientist retract his ideas about the sun's influence on human beings. Any proof that revolution was brought about by anything other than the natural progression of the working class's struggle might prove ruinous to the Communist Party.

When Chizhevsky refused, he was shipped off to a gulag in the Ural Mountains. He spent eight years imprisoned there and, after his release in Kazakhstan, another eight years being rehabilitated, after which, his already fragile health largely broken, he lived only long enough to restore his reputation. A year after his death in the mid-1960s, when the Soviet Academy of Sciences opened his archives, the sheer breadth and foresight of Chizhevsky's work stood fully revealed. He became a posthumous hero, with a science center created in his name, which houses, in pride of place, the Chizhevsky "chandelier," an early air ionizer.

Nevertheless it required years and the work of many other scientists around the world to determine what exactly Chizhevsky had been up to.[1] In America the economist Edward Dewey was one of several to take up Chizhevsky's mantle, largely as an explanation for economic boom-and-bust cycles, in a doomed attempt to help President Herbert Hoover avoid being blamed for the Great Depression. It was not until the 1970s, with the work of the biologist and physician Franz Halberg and his lifelong colleague, the Belgian physicist Germaine Cornélissen, that mainstream science finally began to understand the extent of our reliance upon the volatile activity of the sun.

Halberg and Cornélissen are cycle hunters, both expert in what they call "chronomes," repeating patterns in biology. Today a grand old man in his nineties, Halberg has devoted his working life to studying the effect of external environmental influences on living things. In 1972 he enlisted Cornélissen as a young postgraduate precisely because her doctoral dissertation had been on time-series analysis, which looks for the patterns in events that are repeated over regular intervals. Halberg had already discovered that virtually every one of our biological processes operates according to a predictable timetable.[2] It was Halberg who first coined the term "circadian" (*circa* = about; *dia* = day) for daily biological rhythms, such as our own sleep-wake cycle, and eventually the term "chronobiology," the recurrent cycles of biological function. He also set up the most respected institution to study this phenomenon, the Chronobiology Laboratories at the University of Minnesota, now world-renowned.

Everywhere Halberg and Cornélissen look they see new cycles and periodicities. During years of meticulous research the pair have discovered that the biological processes of every living thing not only have daily rhythms, but also biweekly, weekly, and even yearly cycles; human pulse and blood pressure, body temperature and blood clotting, circulation of lymphocytes and hormonal cycles, variability of heart rate, and most other functions of the human body all ebb and flow according to a relatively predictable timetable. Halberg has demonstrated that blood-pressure level in most people appears to peak between midday and 4 p.m.[3] Even drug treatment such as chemotherapy varies in effectiveness according to the time it is administered.[4]

After puzzling for years over the mechanism for these cycles *within* organisms and whether "time genes" turned on by external influences could exist, Halberg finally concurred with Chizhevsky's conclusion all those years before: the synchronizer of many of these processes is not built in, but is, as certain of his colleagues referred to it, a *zeitgeber*—his native German for "timekeeper"—some sort of *external* environmental signal that set off and entrained, or synchronized, the biological rhythms of all living systems. Although Halberg himself detested the term as imprecise—many of the rhythms seemed slightly irregular—it wasn't until he was in his eighties that he uncovered evidence that the *zeitgeber* resided in outer space,[5] and that the master switch was not light but, just as Chizhevsky had predicted, solar magnetic fields.

Halberg realized that Chizhevsky's discovery went beyond periodicities and cycles, for he had uncovered a sobering truth about the human condition: we are not completely in charge of our own destiny, particularly our biological destiny. The sphere of influence on our biology does not end with our immediate environment or even the Earth itself, but extends to the outer reaches of the cosmos. As many other colleagues of his around the world now confirm, the metronome of every living thing, setting the tempo for our basic regulatory systems and maintaining us in a state of healthy equilibrium, is the sun. So powerful is this cosmic *zeitgeber* that it may even influence our physical size, our longevity, our mental stability, our propensity to violence, and possibly even what we consider our uniquely individual motivation. Our ultimate environmental Bond, which shapes us and our lives, is with a star 93 million miles away.

The Earth is essentially a giant magnet, its North and South Poles, the two poles of the magnet, surrounded by a donut-shaped magnetic field. This ambient geomagnetic field, or magnetosphere, is in constant flux, affected by the weather, any geological changes on Earth, even the wobble of the Earth in its daily rotation, but most particularly by the extreme changes of weather in space, largely caused by the ferocious activity of the sun.

The benign star responsible for all life on Earth is essentially a cluster of unimaginably hot hydrogen and helium the size of approximately one million Earths, crossed with a layer of unstable magnetic fields. Not surprisingly this mercurial combination results in periodic volcano-style eruptions, propelling gas into space as vortices of concentrated fields, the dark blobs on the sun's surface that we call sunspots, pull apart and reconnect in new arrangements. Despite this potentially anarchic combination, the sun carries out this activity according to a fairly predictable timetable; solar cycles consist of approximately eleven years, during which time sunspots build up, discharge, and begin to wane.

During the waxing stage, as sunspots accumulate, so the sun begins to hurl its gaseous explosions our way: solar flares, electrified, bullet-like high-energy protons, corona mass ejections—a billion tons' worth of gas and magnetic fields with the force of billions of atomic bombs—all made airborne and some aimed toward Earth through the electrified gas of the solar wind, traveling some five million miles per hour. This activity causes extreme geomagnetic storms in space, which, during moments of intense solar activity, create a powerful effect on the Earth's magnetic field. During any given eleven-year solar cycle, we can expect to experience two years' worth of geomagnetic storms severe enough to disrupt portions of the Earth's electrical power, interrupt high-tech communications systems, and disorient spacecraft and satellite navigation systems.

Until recently scientists had discounted the idea that the Earth's faint magnetic field (a thousand times weaker than a standard high school horseshoe magnet) had any effect on basic biological processes, particularly as all living things on Earth are now exposed to much stronger electromagnetic and geomagnetic fields every moment of our modern technologically dependent lives. But the latest discoveries have unveiled that living things have a small window through which subtle geomagnetic and electromagnetic fields, such as those generated by the Earth, rather than the artificial kind generated by technology, have the most profound effect upon all cellular processes in living things.[6] Changes in this faint charge, particularly those of extremely low frequencies (less than 100 Hz), profoundly influence virtually all biological processes in living things—particularly the two major engines of the body, the heart and the brain.

This shouldn't be surprising, remarks Cornélissen. "We know when a geomagnetic storm is coming—through our electrical grids," she says. "Electric circuits react to it, as do the heart, brain, and neurosystem. In fact, the heart is the biggest electrical system of the body." As she views it, a human being is simply one more satellite system prone to being destabilized or even blown off course by an electrical storm in space.[7]

Magnetic fields arise from the flow of ions, the electrons and atoms with charge. When magnetic forces change direction, as they often do on the surface of the sun, they shift the direction of the flow of atoms and particles. All living things, including humans, are made of the same basic material, and as Chizhevsky intuited, any change of magnetic force will alter our own internal atomic and subatomic flow.

The Earth's magnetic activity appears to most directly affect our cell membranes and calcium-ion channels, which are vital for regulating enzyme systems within the cell. In particular the Earth's geomagnetic field appears to target the sympathetic nerves (those originating from the chest and lower-back parts of the spinal cord), including the fight-or-flight response.

Of all the systems in the body, changes in solar geomagnetic conditions most profoundly disturb the activity of the heart. In susceptible people geomagnetic storms can bring on a heart attack.[8] Healthy hearts have a great degree of variation in heart rate, but a good deal of geomagnetic activity decreases heart rate variability,[9] which in turn increases the risk of coronary artery disease and heart attack. When geomagnetic activity increases, the blood gets much thicker, sometimes doubly so, and the bloodstream slows down—a recipe for a heart attack.[10] Heart attack rates and cardiovascular death swell whenever there are increases in solar cycle geomagnetic activity,[11] the largest number of sudden deaths from heart attacks occurring within a day of a geomagnetic storm.[12] In one study Halberg tracked the heart attack rate in Minnesota over a number of years and discovered a 5 percent increase during times of maximum solar activity.[13] The human heart rate also appears to be affected by large changes in the solar wind, becoming amplified especially when the solar wind changes velocity over a seven-day period.[14]

As if to make up for the Soviet persecution of Chizhevsky, Russia has been in the vanguard of research into the effect of solar geomagnetic activity on the body, initially to determine the effects of space weather on cosmonauts while in space. Soviet scientists discovered that a cosmonaut increased his risk of a cardiac arrest during a magnetic storm.[15] Cornélissen also found that the healthiest (that is, the most varied) heart rates of their volunteers on Earth occurred during times of the least amount of solar activity,[16] whereas certain electrical frequencies were dampened down and heart rate variability decreased during magnetic storms.

Besides its effects on the heart, the sun has a profound effect on the other electrical center of the body: the brain and nervous system. Scientists in the former Soviet bloc have discovered that even in healthy volunteers electrical activity in the brain gets highly destabilized during magnetically stormy days.[17] Solar activity also corrupts the very signaling within the nervous system, so that some parts are overactivated and others fail to fire.[18] Scientists at the Azerbaijan National Academy of Sciences at Baku showed that big upsets in geomagnetic activity appear to disturb the balance of the electrical communication system of the brain, overexciting certain parts of the autonomic nervous system and lowering the activity of others.[19]

When the sun explodes, so, in a sense, do we. Geomagnetic activity in space upsets our energetic equilibrium, profoundly affecting our mental stability. During a magnetic storm the mentally disturbed get even more so. The higher the geomagnetic activity, the greater the increases in general psychiatric disorders, the greater the number of patients hospitalized for nervous conditions, and the greater the number of attempted suicides.[20] An American orthopedist, Robert O. Becker, who carried out numerous experiments on the effects of electromagnetic fields on health, found a link between violent solar storms and admissions to mental hospitals.[21]

Several years ago Halberg and Cornélissen collaborated with a number of neuroscientists to determine whether autism might be influenced by geomagnetic factors. Up until that time no seasonal variations in autism had been detected; no more children develop autism in winter, say, than in spring. When Halberg and his colleagues compared the incidence of autism with solar activity, however, they found a distinctive signature between geomagnetic cycles of 1.9 years in length and failure of children to bond with their mother. Solar influences were so strong that they affected maternal love.[22]

There is even some research, particularly from Canada's Laurentian University, showing that epileptic fits result from geomagnetic disturbance or are exacerbated by them.[23] Sudden death from epilepsy and sudden infant death syndrome have also been linked to high levels of geomagnetic activity.[24] In one study researchers found that geomagnetic activity affecting the Earth's geomagnetic field was significantly stronger on days that patients suffered seizures.[25]

Cornélissen's own specialty is the unseen cycles, relating to solar wind and the changes in the sun's position during a change of season, such as during equinoxes, when the sun appears in the same plane as the Earth's equator, and day and night are approximately equal in length. She has found a vast array of solar-cycle signatures in mental illness and epilepsy: epilepsy is more pronounced during the vernal (spring) equinoxes, and suicides and depression follow a 1.3-year cycle, which corresponds to the cycle of the solar wind and interplanetary magnetic fields. Even the number of traffic accidents regularly rises and falls in time with the solar seasons.[26]

Other evidence has bolstered her research. A study by scientists at the University of Melbourne compared all the suicide statistics for Australia from 1968 to 2002 with the daily index of geomagnetic activity for those years. They discovered a fascinating difference between the sexes in their figures: there were pronounced seasonal fluctuations corresponding to solar activity in the times men were most likely to kill themselves, but women appeared to have the greater solar connection and were far more likely to kill themselves during times of strong solar flares (which occur every five months) or solar wind.[27]

Even our basic statistics at birth—our birth weight, height, and head, chest, and abdominal circumference—believes Halberg, are linked to the ebb and flow of Hale solar cycles (the sun's complete cycle, presently about twenty-two years): the greater the solar activity around the time of birth, the larger the child.

As a physician Halberg's main interest in chronobiology is medical. As he sees it, our seeming powerlessness in the face of an outside *zeitgeber* is a source of comfort; if patterns like this can be predicted, we can create compensatory behavior. Heart infections following a heart attack, for instance, are known to increase during times of geomagnetic disturbance, so patients at risk can be given antibiotics as a preventive measure.[28]

To that end Halberg and Cornélissen created BIOCOS (BIOsphere and COSmos), a vast multicenter project that aims to act as an early warning system by continuously monitoring certain physiological variables caused in susceptible people by the sun and the planets. The Phoenix Ambulatory Blood Pressure Monitor Project in the Twin Cities of Minnesota, for instance, intends to offer volunteers an ambulatory blood pressure monitor so that those with high blood pressure can track effects on their arteries when solar activity is at its peak. The project specifically targets reduced heart rate variability and a phenomenon called CHAT (Circadian Hyper-Amplitude Tension), or excessively large swings in blood pressure during one part of the day. During high levels of geomagnetic activity patients with heart conditions are advised not to exert themselves suddenly. BIOCOS also plans to investigate techniques that shield against natural geomagnetic disturbances.

Although gratified by all this progress, Halberg despairs that it will take many more years, well beyond his own ebbing life span, before American medicine accepts the notion that the weather in space powerfully influences our biology. Russia is one of the only countries to make a serious attempt at this kind of preventive medicine. For instance, at Georgian Technical University three sets of Helmholtz coils bathe a room in a powerful magnetic field. Yuri Gurfinkel and his colleagues intend to use this setup for cardiovascular patients, particularly those in intensive care. The coils will act as a powerful form of preventive medicine, offering a compensatory shield against what Gurfinkel regards as a more potent killer than diet, lifestyle, or even genes.[29] By creating these early warning systems and geomagnetic shields, these scientists openly acknowledge that our physical health—even our physical size—is utterly dependent on solar whim. But what has also come to light is how much our behavior is also dependent upon our Bond with the sun.

In his twilight years Halberg has turned his attention to determining whether there is any truth in Chizhevsky's claim that the sun has influence on the capriciousness of the human heart. He and Cornélissen received a windfall in the form of worldwide data on virtually fifty years of activity by Jehovah's Witnesses, a painstaking record of the hours each member from 103 geographical locations spent soliciting on behalf of their churches. Because every Jehovah's Witness is duty-bound by the church to attempt to recruit new members, a record of their activity offered Halberg and Cornélissen a unique opportunity to study whether their efforts correlated in any way with solar activity.

When the pair mapped these data on a graph, they found a giant peak and trough in recruitment efforts over a twenty-one-year period, which directly corresponded with the peak and trough of the sun's twenty-one-year Hale cycle. When they then looked more closely at the data, comparing the activity of church congregations from different locations with solar activity in corresponding latitudes, they discovered that these also were associated with rises and falls in geomagnetic activity. This work offers compelling evidence that geomagnetic activity affects certain brain areas involved in motivation, just as it affects physical function, body measurement, and development.[30]

Other researchers have been inspired by Halberg's example to revisit whether a population's tendency to spend or save is also affected by the sun, as Dewey had claimed during the Great Depression. Earlier research had shown that geomagnetic storms profoundly affect people's mood, which is related to judgments and decisions concerning risk. Naturally one area of immediate interest for banks is the potential effect of solar activity on the stock market, for if a lending institution can predict what is essentially a betting gamble, they stand to make a killing. To examine this further, the Federal Reserve Bank of Atlanta recently teamed up with Boston College to study the buying and selling habits of Americans during geomagnetic cycles. They discovered that, during geomagnetic storms, people were more inclined to sell stocks; they tended to misinterpret their own negative physical response to solar activity as external evidence of a negative economic climate. Consequently demand for riskless assets vastly increased, causing the price of riskier ventures to tumble or rise more slowly than otherwise.

After controlling for market seasonal cycles and other types of environmental and behavioral factors known to affect the market, the Bank of Atlanta research team concluded that geomagnetic storms have a negative effect on the following week's stock returns for all U.S. stock market indices. Periods of solar calm, on the other hand, show evidence of substantially higher returns.[31]

The Technical Securities Analysts Association of San Francisco went a step further to determine whether this activity governs financial boom-and-bust crises, particularly as it is well known that a collective mood governs whether people are likely to engage in buying frenzies or bearish selling. The analysts discovered that financial crises follow a fifty-six-year cycle, which itself follows a predictable cycle of the moon and sun, when angles (from 0 to 180 degrees) between the sun and moon are repeated to within a single degree.[32]

After the 9/11 attacks Halberg and an international BIOCOS team of Americans and Russians turned their attention to terrorism, comparing solar activity with the timing of international acts of terrorism for forty years, from 1968 to 2008. They discovered that peaks of terrorism matched the cycles of solar wind and the Earth's geomagnetic index.[33]

Ninety years after Chizhevsky was sent to the gulag for his crazy theories, Halberg worked with the Russians to prove that Chizhevsky may have been onto something after all. Halberg and his team of international colleagues now understand that biology and behavior are not wholly individual, and that all living things resonate in every way with the cosmos.

In the early hours of July 29, 1976, an overweight, cherub-faced eighteen-year-old emerged from his yellow Ford Galaxy compact, produced a .44 caliber Charter Arms Bulldog revolver from a paper sack, crouched, and took pointblank aim at Donna Lauria and Jody Valenti, two teenage girls who had been sitting in Valenti's parked Oldsmobile in Pelham Bay after a night at the Peachtree discotheque. Valenti was wounded, but Lauria died instantly.

The New York Police Department's Eighth Precinct assumed they were dealing with a spurned lover or a bungled mob assassination attempt. After three more Queens shootings the following October, November, and January, however, with bullet slugs found to have the same distinctive marks, the police realized they had a serial killer on their hands.

The city's police force and the media began to catalogue similarities in the modus operandi of the crimes in the six subsequent attacks, particularly after Son of Sam, as the papers dubbed the killer, provided clues in two letters, one left at the crime scene for Police Captain Joseph Borrelli, and the other later posted to the *New York Daily News* columnist Jimmy Breslin—similarities they continued to revise. Sam zeroed in on young couples in parked cars. Sam only struck on weekends in the early hours. Sam's hunting ground was lovers' lanes in the Bronx and Queens. Sam hated the city's current disco fever, following the recent huge success of the film *Saturday Night Fever,* and was targeting couples leaving discotheques. Sam had a penchant, it seemed, for young women with long dark hair, causing a run on hair cuts and blonde wigs—until he shot a blonde two days after the first anniversary of the killings. Now it seemed that Sam's choices were random, uncontrolled. "NO ONE SAFE FROM SON OF SAM," announced the front page of the *New York Post* on August 1. In one of the hottest summers on record, people huddled behind drawn curtains at night, and everyone stayed out of parked cars.

What remained unremarked upon was the one pattern of Sam's MO that was more or less consistent: five of the eight attacks and all but one of the murders had occurred during a full or new moon.

After the murderer, David Berkowitz, was caught, confessed to the killings, and began serving six life sentences, some members of the police and writers such as Maury Terry, the author of *The Ultimate Evil*, grew convinced that Berkowitz may have been involved in a cult that chose the time of the killings on purpose, as part of some Satanic ritual.

Even if Berkowitz acted alone, most policemen would not have been surprised that he struck during special times of the lunar cycle. As any cop on the beat believes, a full or new moon brings out the worst in people. Police gird themselves for a higher crime rate and more outrageous phone calls than usual, psychiatric hospitals for higher admission rates, hospitals for a higher number of casualties, and teachers for more unruly classrooms.

The common belief is that the moon's effect destabilizes people so that they are more violent during certain times of the lunar cycle. Homicides, road accidents, accidental poisonings, suicides, and emergency room visits are all thought to increase during full and new moons. In one study Arnold Lieber, a Miami psychologist who compared the timing of homicides with lunar activity over a fifteen-year period, found that the number of killings in Dade County rose in remarkable tandem with the full or new moon, tailing off significantly in the first and last quarters.[34] Even animals are thought to bite more often when the moon is full, according to one British city's accident and emergency department data from 1997 to 1999.[35]

Psychiatric effects are supposed to follow the opposite ebb and flow—highest during a new moon and lowest when the moon is full. In one eleven-year study of nearly 19,000 patients in a psychiatric hospital, the incidence of patient psychosis peaked during a new moon and was at its lowest during a full moon.[36] Suicides followed suit: among emergency calls to suicide-prevention centers tracked over a two-year period, the highest number for women occurred during the new moon, not the full moon.[37] The so-called lunar syndrome is even thought to affect workers' absenteeism; studies have shown that many more individuals visit the doctor during a full moon than at other times.[38]

Not all studies make such a neat association, however, and one prob-

lem with the data is that the researchers are looking for a simple relationship (such as an effect only with the full moon), whereas the truth
may well be far more complex.

The received wisdom has been that any lunar influence is due to the
gravitational effect of the sun and moon, as it is with the tides; that
is, because we are 75 percent water, the moon affects us just as it does
the ocean. However, the tides are predictable, occurring every twelve
hours, whereas lunar effects are seen only once or twice a month.

The most likely explanation, says Cornélissen, is a subtle geomagnetic effect, or some influence of the moon on the sun's well-established
geomagnetic pull. During a full moon the Earth sits between the moon
and the sun so that both enter our geomagnetic field. During a new
moon the places are reversed: the moon sits between the sun and the
Earth and is farthest away from our geomagnetic field. It is likely that
the placement of the moon actually amplifies or muffles the geomagnetic pull of the sun and the Earth's geomagnetic field. It is also helpful
to remember that the lunar synodic month (29.5 days) is only a few
days longer than the period of the full rotation of the sun.

All this makes sense if you regard the moon as a giant magnet that
modifies solar geomagnetic influence. In fact it may physically be a
magnet. Studies of lunar samples brought back from the Apollo flights
show evidence of strong magnetic fields in the rock.[39] If the moon has
its own strong magnetic pull, this could cause a magnetic shift when
the moon passes through the Earth's geomagnetic tail, as happens at
every new moon.

The sun and the moon are not the only celestial bodies with an effect
on our physical and mental activity. Several years ago Jerry Mitrovica,
a physicist at the University of Toronto, and Alessandro Forte of the
Institut de Physique du Globe de Paris published a paper in the prestigious scientific journal *Nature* showing, through mathematical calculations and simulations, a relationship between tiny changes in the
Earth's shape and axial rotation, and the gravitational effects of other
planets in our solar system, particularly Jupiter and Saturn.[40]

The paper was fairly technical, but beneath all the scientific jargon
the physicists were making big claims. "We're showing for the first time
that changes in the Earth's shape, when coupled with the gravitational
effects from other planets, can produce large changes in the Earth's

climate," said Mitrovica, who is working on behalf of the Natural Sciences and Engineering Research Council and the Canadian Institute for Advanced Research (Earth Systems Evolution).[41]

In his mathematical model Mitrovica demonstrates that the Earth's orbit is affected by the gravitational pull of Saturn and Jupiter. At some point during the past 20 million years the Earth encountered gravitational "resonance" with the orbits of Jupiter and Saturn, he says, which ultimately influenced the angle of tilt of the Earth's axis during that period.

The gravitational pull of any particular planet is extraordinarily small, and many scientists don't believe that, on its own, it would have much of an effect on the Earth's geomagnetic field. However, Halberg, Cornélissen, and their colleagues believe that a planet's pull can create "tidal" effects, in which the gravitational forces of the various planets also interact with the magnetic fields of the sun and moon as well as the solar wind. This would have a cumulative effect on the magnetosphere, which ultimately can have profound effects on both climate and biology. Slovak researchers have carried out work on these "lunisolar tidal waves" and have demonstrated that the geomagnetic activity, as with solar activity, correlates with the incidence of illnesses such as epilepsy.[42]

But the story of planetary effects may be even more complex: all planets may exert gravitational effects on each other, which is likely to result in a nonlinear, or chaotic, effect. Carl Murray, an astronomer at Queen Mary College, University of London, has carried out research showing that planets' orbits are elliptical and planets rotate on their axes with a particular degree of tilt because of gravitational effects.[43] A resonance effect can also be established between two celestial bodies when the time periods of their rotations around each other lock into a regular mathematical relationship. For instance, the moon rotates around the Earth for the same amount of time as it rotates on its own axis. Other bodies in the solar system may circle around each other at a regular interval, say, at two to three times what it takes them to rotate on their own axes. Although the relationships between planets can slow down or speed up a rotation only slightly, even the subtlest of changes may have a large effect on weather and biological life. These kinds of gravitational effects are magnified when a variety of planets are in alignment, as occurs with the sun, the moon, and the Earth during an eclipse.

Besides the chaotic effects of gravity, the electromagnetic fields created by each solar body can interact with each other and affect the sun, the moon, and, of course, the Earth. Indeed some scientists believe that the influence of planetary fields from the Earth and other planets triggers solar activity like sunspots, and not the reverse. When the Earth and the sun are positioned at a particular angle to one of the major planets, such as Saturn or Jupiter, this too will affect the formation of sunspots or bursts of solar plasma. It is also known that the interplanetary magnetic field between the Earth and the sun and the Earth's geomagnetic envelope interact more during the equinoxes, largely due to the Earth's spin on its axis.

Scientists have long known that when planets are at major angles to each other in relation to the sun (at 90 or 180 degrees, for instance) they will affect the reception of radio signals—the same kind of effects, albeit in a more minor way, caused by unsettled geomagnetic solar activity. These subtle interrelationships could all add up to large effects on Earth.

All this may sound extraordinarily dense and complex, and a bit like a scientific version of astrology, but it is not difficult to understand if we alter our perception of what we are: part of a larger planetary system. "To understand climate on Earth, it's clear that we need to consider the Earth as this dynamic deforming system," Jerry Mitrovica says. "But we also need to understand, more than we thought we did, the Earth's place in the solar system."[44]

We have to develop greater appreciation of the fact that we live within a cosmic Bond of complex interrelationships and constant flux. Rather than discrete entities, living things and the Earth itself are part of an energetic system dependent upon other outer forces, gravitational and geomagnetic. Halberg regards this effect poetically. The living organism, he says, must be viewed as "a dynamo and a magnet, living on the Earth, a larger magnet, in the atmosphere of the sun . . . with magnetic storms causing blackouts in cities and . . . in human hearts."[45]

The importance of Chizhevsky's discovery and Halberg's evidence cannot be overestimated. If we are essentially at the mercy of the slightest move of the sun and its activity, their work stands as a giant refutation of our misplaced belief in ourselves as masters of the universe—or even of ourselves. The Earth, its inhabitants, and all the other planets around us exist within a sphere of collective influence, resonating in unison. Our true *zeitgeber* is the collective effect of all the solar system.

THE BOND

Ultimately it is difficult to consider our universe as anything other than a unified whole. We can begin to take charge of our own destiny only when we consider the Bond in its entirety, as a superorganism, completely interrelated.

In a manner of speaking, we already do. It is actually through our interdependence that we learn to comprehend our world.

CHAPTER 4

SHARED CIRCUITRY

In 1991 Giacomo Rizzolatti, a neurologist at Italy's University of Parma, took delivery of an expensive new piece of equipment for his laboratory, which, in one sense, was nothing more than a new toy for his monkeys. His current work required a good deal of playing with his little cadre of pig-tailed macaques, and, as he saw it, his team's primary challenge was keeping the monkeys constantly amused. Crucial to the game was maintaining a certain element of surprise. With that in mind, Rizzolatti had a $60,000 apparatus constructed consisting of a black box with a rotating wheel inside that was powered by electricity, most of which was encased in spherical siding. When turned on, the wheel would rotate at random until one of the prizes on a small shelf—a cube, a pyramid, a sphere—would suddenly be revealed, at which point the entire mechanism would illuminate, as if accompanied by a drum roll.

The monkeys had been trained to remain motionless until an LED light turned green, at which point they were free to snatch the prize. The gadget also had to capture a monkey's attention in order to distract it from the real point of the exercise. A micro-electrode had been attached to its scalp, which, in turn, was connected to an oscilloscope—a visual display showing the waves of electronic signals—and an amplifier, in order to record every firing of a single neuron. The two types of equipment captured the activity of the neuron from two perspectives, as a visual display and also as a sound. The amplifier was particularly useful; by sounding a noise every time the neuron fired, the amplifier enabled the scientists to play with the monkeys while taking note every time the sole neuron discharged.

Rizzolatti, whose shock of wild hair gave him the odd countenance of an Italianate Einstein, was a specialist in movement, the body's motor system and its role in cognitive function. He, a team of post-doctoral students—Vittorio Gallese, Leonardo Fogassi, and Luciano Fadiga—and two formidable colleagues, the noted French neurophysiologist Marc Jeannerod and Michael Arbib, a British biologist and then a professor of computer science in America, periodically met in Rizzolatti's lab to pursue a seemingly modest goal: to isolate the exact sequence in the brain through which a living thing transforms visual information into action. By working with the monkeys the team hoped to discover which neurons in the brain are responsible for controlling the motor sequence of particular hand and mouth actions during the tiny window of time between seeing an object and reaching to grasp it. They were studying the ventral premotor cortex area, the portion of the brain responsible for formulating and carrying out intentions, in the certainty that the brain must engage in forward planning before the arm decides to reach out and take hold of something. Despite appearances, it was a painstaking operation, requiring that Rizzolatti's team take note of the firing of the neuron every time the monkey reached for the food.

On the day they were due to introduce one of the macaques to the Black Box, as they'd begun to call it, one of the researchers was reaching to plant one of the objects into the spherical center when the beep of the amplifier sounded, signaling that the neuron of the monkey sitting in the chair opposite him had fired. The neuron in question was one Rizzolatti knew to be involved only with movement, but as they hadn't yet got to the part of the experiment where the monkey was to reach out for the object, he and his fellow researchers discounted the effect, figuring the monkey must have moved his hand or finger. For a number of months the beep sounded during the setup stage of nearly every experiment, and each time the team dismissed this as either an artifact—an error caused by the equipment itself—or an indication that the monkey was restless, moving his arm in impatience, or making a premature attempt to snatch the prize.

After some months Rizzolatti could no longer ignore this strange phenomenon. It seemed impossible that the monkey would move almost every single time before it had any motivation to do so. Rizzolatti decided to embark on further tests to confirm that the monkey's body was indeed shifting in any way. His team began to record discharges from the monkey's primary motor cortex, which would readily identify the slightest movement, and also made use of electromyography, which records every electrical impulse sent from motor neurons to specific muscles to cause a contraction. Neither piece of equipment showed any evidence that movement had occurred.

Rizzolatti was thunderstruck when it finally dawned on him what must be happening: the *very same neuron* in the monkey's brain that fired when it intended to grasp the object in the Black Box was also firing when the monkey *observed* the researcher grasping it.[1]

Rizzolatti began testing his hypothesis by examining other neurons involved in movement. The more he looked for these strange copycat neurons, the more of them he seemed to find. More than three-quarters of the neurons used for a specific function of movement were carrying out double duty, when the monkey was simply observing the same movement made by a human.[2] It fascinated Rizzolatti that the monkey did not require a mirror image of itself in order to internalize the observed activity; its neurons also fired when the monkey was observing an action by a human, another species altogether.[3] Nevertheless when it came to actions and intentions, as Rizzolatti learned in subsequent research, these copycats were highly specific. A particular neuron would fire when a researcher was carrying out a specific action with clear intention and a goal that was comprehensible to the monkey, such as grabbing for an apple and placing it in his mouth, but nothing fired when the monkey observed an action that he didn't recognize or could not relate to, such as placing the apple in a cup.[4] It appeared that the intention had to be within the monkey's own movement repertoire. These neurons also appeared to be audiovisual, activated not only by the sight of movement, but also the *sound* of it, particularly when the monkey could not see what was happening.[5] This confirmed that the purpose of these neurons was to aid in understanding the intentions of others so that the monkeys could anticipate what was likely to happen next.

The effect occurred not only in the lower portion of the frontal lobes but also in the posterior parietal cortex of the brain. This section of the brain both helps to make sense of disparate sensory information and to distinguish *self* from nonself. Although the monkey's brain should know that observed action is different from performed action, it wasn't acting as though it did.

*   *   *

Rizzolatti quickly forged ahead with his first research on humans, working with Scott Grafton, a neurologist at the University of Southern California, whose laboratory had the most advanced brain-imaging equipment at the time. They were not able to locate individual mirror neurons with the precision they'd been able to achieve with the monkeys, largely because ethical review committees will not allow researchers to drill holes in human skulls—a necessary part of attaching single electrodes. But they could call upon the next-best thing: a brain-imaging technique that showed the general part of the brain and the system of neurons that were being activated.

At the time of their early experiments the only imaging technique available to the scientists was positron emission tomography, which showed certain areas of brain activity, but without much precision. As time went on and brain-imaging technology became more sophisticated, Rizzolatti was able to use functional magnetic resonance imaging (fMRI), which records even the tiniest of changes in brain blood flow, and transcranial magnetic stimulation, which measures the size of discharges in the motor cortex and consequently can pinpoint neuronal activity more precisely.

But even with the crudest equipment, Rizzolatti discovered the same phenomenon in humans that he had observed in monkeys: an identical, finely orchestrated, and highly specific system of brain and spinal neuronal activity that occurred whether their human subjects were performing an action or observing it.[6]

The copycat phenomenon posed a great mystery. Rizzolatti first assumed that these neurons fired so that a living thing could learn how to do something quickly in order to survive. He thought it might be part of "Hebbian learning," identified by Donald Hebb, who suggested in 1949 that neurons become more efficient and operate as a unit when they repeatedly and persistently are stimulated together. *Neurons that fire together wire together.*

Rizzolatti also wondered whether the activity might be involved in helping a monkey to learn through imitation. But why would an adult monkey need these neurons as a learning tool? All evidence suggested that this breed didn't learn by copying.[7] He consulted some primatologists, but they poured water on the idea: *Chimps imitate; monkeys don't imitate.* Only a tiny window of time was open to baby macaque monkeys for learning by copying gestures. Furthermore, as Rizzolatti well knew, both newborn humans and monkeys are immediately able to copy complex facial actions without instruction or practice.[8] The most often cited case is when a mother sticks out her tongue to her newborn baby and the baby copies the gesture immediately, even though it is a complex maneuver requiring the delicate coordination of a host of neuronal sequences. Even baby macaques are able to imitate the tongue-sticking-out exercise.

Rizzolatti concluded that within the brain of both primates and humans there must be no distinction between observation and action. *To make sense of the riot of experience around us we can only imagine it by mentally undergoing it ourselves.* We understand the actions of others by simulating the entire experience from a personal vantage point, *as though it were happening to us.*

Rizzolatti realized that they had uncovered something fundamental in the neurobiology of comprehension. He began referring to the phenomenon as a "mirror effect," because the neurons had a dual purpose: to impel the muscles to act and to take note of action in others.

Convinced that they had uncovered an important means by which the brain connects with the outside world, Rizzolatti and his team wrote up a modest paper detailing a year of their research and sent it off to the prestigious science journal *Nature,* only to have it rejected out of hand as not sufficiently interesting to anyone outside of neuroscience. Through a personal connection, Rizzolatti eventually managed to publish the paper in the *Journal of Experimental Brain Research.*[9] Five years later, after the paper was widely read and its full implications digested around the world, *Brain,* the most important journal of neurology, welcomed an updated version of their original research and published it immediately.[10]

Among Rizzolatti's team of postdoctoral students was a young German researcher named Christian Keysers, who'd just arrived to work in the Italian lab from St Andrews University. As Rizzolatti and Keysers went on to discover, human beings make use of mirror neurons to read emotion as well as action. The same sections of our brain that activate when we experience a range of human emotions, from joy to pain, also activate when we observe the emotion in another.[11] We simply need to observe a person's facial expression or body language to set the cascade of neurons in motion.[12] When you see someone smile or grimace, as far as your brain is concerned it is you who are happy or disturbed.[13]

In a study that would make his name, Keysers monitored the brain activity of a group of participants while they were inhaling repulsive or pleasant odors then watched a film of individuals reacting to similar odors. He discovered that the same portion of the brain, the anterior insula, was activated in the participants whether they were breathing in the smells themselves or simply observing someone else's facial expression as that person reacted to the odor.[14]

Mirror neurons are meant to work out not only *what* someone is doing and *how* he feels about it, but also *why* he is doing it. Rizzolatti discovered that the neurons do not fire if the goal of the action is unclear. In one study a human observer's mirror neurons lit up when observing the actions of a robot, but only when the robot was carrying out a clearly defined task,[15] not when the robot was repeating the same task over and over.

Rizzolatti's discovery has now been universally acknowledged as a remarkable advance in our understanding of how the brain processes the actions and emotions of others. Less recognized, however, are the enormous implications of his discovery on the biology of perception and social interaction. As his work makes clear, perceiving the world is not an individual affair, limited to our own mental capacities, but a process involving shared neural circuitry. We internalize the experience of others at every moment, automatically and immediately, without conscious effort, using a neural shorthand created of our own experience. In the very act of connecting with someone, even on the most superficial level, we are involved in a relationship of the utmost intimacy. Our understanding of the complexities of our world occurs through the constant melding of the observer with the observed.

Keysers realizes that there are always two vantage points in the act of perception. "During most social interactions there is not a single agent and a single observer," he wrote. "Both partners are both observer and agent, both the source and the target of the social contagion the mirror neuron system conveys."[16] To watch someone is to immediately internalize his point of view. This means that the very act of observing another person automatically engages us in a Bond in which we, the subject, fuse with our object. In a manner of speaking, in order to understand another person, we must temporarily merge with him.

In spite of our brain's great capacity in every other regard, our method of taking in all we see around us, particularly the activity of other living things, is highly unimaginative. When we observe the action of another person, in order to make sense of it, we have to re-create the experience in our head *as if we were carrying it out ourselves.* We translate the actions, sensations, and even the emotions of others into the neural language of our own body, as though it was our own experience.[17] The same neurons fire in our heads whether we feel the touch of something against our legs or observe something touch someone else's legs, or even watch an object being touched. Any type of touch we see evokes the neural networks involved in our own subjective experience of touch.[18]

As is the case with monkeys, Rizzolatti realized, for a human mirror neuron to activate, the observed activity has to be within the observer's own motor repertoire and derive from his experience. For instance, we can immediately connect with a dog through our mirror neurons when we see it eating a succulent piece of meat, but not when it is barking at another dog. Without the ability to "feel into" the experience through mirror neurons, the brain of a human can create only an approximate experience by cobbling together a rough simulation, much as a computer would, of what it is to bark.

In fact we always filter our observations of someone's activities through our first-hand experience, even if it differs from that of the person we're observing. Keysers once studied the brain activity of a person born without hands who was observing another person reaching for a glass. In this instance the regions of the brain and spine associated with the toes and feet were activated, not the areas associated with hands. The handless observer understood the action through the same process that he used to grasp a glass: with his feet.[19] This would suggest that the act of seeing establishes a Bond—a complex mix of your actions and emotions and mine.

If you were able to get inside your own head and observe your brain and nervous system in the act of relating to someone else, you'd be hard-pressed to figure out which instructions relate to you and which to the other person. You may think that you are an objective observer, but you are always looking through someone else's eyes. The boundaries between you and everyone else are blurred, because they are governed by a complex mix of neural firings originating from inside and outside your head. With no conscious effort you internally re-create the actions and emotions of others through the complex filter of your own experience. If you are speaking to me, your emotions flicker in first, but then I add my own history to the brew.

We not only copy the motor program of the particular action; we also replicate all the physical and emotional feelings associated with it, according to our past experiences, such as whether an activity was hard on the muscles or tingled against the skin. If we're watching an athlete during sports training and we ourselves hated running at school, our old emotions get poured into the mix of our observation. We understand the experience through the link with our own history.

In fact the more familiar we are with the actions we observe, the more of our mirror neurons fire. In a professional dancer observing another dancer, for instance, more of her mirror circuitry involving dance moves are activated than in someone who is less familiar with dance.[20] Every time we look outward we pick up the living world's experience and add to it like someone adding her own favorite ingredients to a new recipe. This makes for a global interpretation of what we've just seen, because, for an instant, we observe the world from a higher perspective—from multiple vantage points and through all the time we've been alive.

This automatic tracking system illustrates the beautiful economy of the body; translating what others do and feel into our own actions and feelings enables us instantly to understand someone else's experience. This is likely how we intuit a person's interior life and empathize with actors in a film or play.[21] Watching a tarantula crawl across James Bond's chest literally makes our own flesh crawl, because, in a sense, we are experiencing the physical sensations of a tarantula crawling across our own chest as well as all the emotions that go with it.[22] Our hearts pound when the bad guy chases our hero, we wince if he gets shot, we're elated when he prevails *because it all, in a sense, is happening to us.*[23] A group of Israeli scientists were able to reconstruct the correct sequence of violent scenes in a gory action movie simply by studying the brain-imaging records of a group of participants who had watched it.[24]

When we say to someone, "I feel your pain," it is actually true. Mirror neurons associated with pain also fire when we see someone getting hurt. In one study participants whose brain patterns were being monitored were asked to imagine receiving a pinprick and then to watch someone else receive a pinprick. The scientists conducting the study discovered that the same neurons fired whether the participants imagined themselves getting pricked or witnessed someone else getting the same injury.[25] However, our ability to feel someone else's pain appears to be dependent on the emotional context of the pain.[26] The neurons that fire create our reaction to pain, not the physical aspects of it. We simulate the emotional experience, rather than actual physical hurt.[27]

Even when you see an enemy in pain, although you may get some perverse satisfaction from the situation, your very first response is a pure connection by putting yourself in the same emotional state. "Initially, you have to understand that this guy is in pain," says Rizzolatti, "and to do that you feel the exact same emotional pain that he does."[28] The act of perception is a moment of perfect union—no matter with whom.

Many psychologists and neuroscientists now believe that mirror neurons represent the first glimmers of empathy—our ability to place ourselves in someone else's shoes—and this appears to be a finely tuned feedback system. Those who rate themselves as highly empathetic typically display more mirror neuron activity.[29] Conversely, as we develop empathy our mirror neuron circuitry develops in complexity, suggesting that the heart of empathy is embodied simulation.

But the reverse is also true: the more finely attuned a person's mirror neurons are, the more likely he is to show empathy. The neuroscientist Antonio Damasio used brain imaging to determine which areas of the brain lit up in a group of participants when they were asked to think about one of three scenarios: an emotional experience from their own past; the equivalent experience for another person, but imagined by the participants as if it were happening to themselves; or an unemotional experience from their own past. When a participant related strongly to another, he produced brain activation equivalent to what would happen if he himself were undergoing the experience. However, when he couldn't empathize with the other person's story, unrelated portions of his brain lit up.[30]

The human brain is a relentless copycat from the moment it is born. Our brain's first impulse is to merge with the brain of our mother. The American neurologist Allan Schore, who has done seminal work on attachment theory, believes that a baby's neuronal system learns from his mother's brain, which acts as a kind of brain-wave template, teaching the baby's brain when to fire and wire, just as she teaches him to talk or use a spoon. Eventually, Schore says, "the prefrontal cortex of the mother becomes the prefrontal cortex of the infant."[31]

Researchers at the University of Arizona found evidence that the brain patterns of a mother, as recorded on an electroencephalograph (EEG), are encoded within the EEG pattern of her child.[32] As the science author Joseph Chilton Pearce discovered, mother and baby usually experience brain-wave entrainment, in which the electrical waves of the two brains "resonate," peaking and troughing at the same time, when the two are together. When one is separated from the other, the brain waves of both become discordant and return to resonance only when the pair are reunited.[33]

Throughout your life your brain finds other brain waves contagious. As I recounted in my book *The Intention Experiment* a great deal of evidence shows that under many types of circumstances, particularly when two people do things together with a singular purpose, the electrical signaling in their brains quickly synchronizes.[34] The frequencies, amplitudes, and peaking and troughing of the brain waves all begin to resonate in tandem.

The research examining this has historically involved some variation of a sender and receiver pair, in which both are isolated in separate rooms, wired up to a variety of physiological monitoring equipment, such as EEG machines. One of the partners who is stimulated with something—a picture, a light, or a mild electric shock—attempts to transmit the mental image of the stimulus to his partner. In a significant number of cases the receiver's brain waves begin to imitate those of the sender when he is being stimulated; the receiver's brain picks up and mimics the experience of his partner.[35] In fact the response of the receiver occurs in a place in the brain identical to that of the sender.[36]

This kind of bonding can occur instantly, even between strangers. Simply being partnered with someone establishes minds in synchrony.[37]

This kind of entrainment isn't limited to the brain. In a remarkable series of studies researchers at the Institute of Noetic Sciences in Petaluma, California, discovered that when one member of a couple sent healing thoughts and intentions to his partner with cancer, a large number of physiological processes—heart waves, brain waves, conduction of electrical impulses in the fingertips, blood flow, respiration—began to mimic each other in both partners. With a loving thought, two bodies quickly became one.[38] Further research at IONS demonstrates that we also register the emotional states of other people in our gut.[39]

Forms of bodily entrainment appear to occur even among strangers, under certain circumstances. Connecting with another person in any intimate way, as a healer does when he sends healing energy to his patient, causes the two brains to synchronize.[40] Even touching someone with a sense of appreciation and kindness can cause that person's brain waves to entrain with yours.[41] Synchrony can even occur when two people have a strong wish to cause harm, as has been demonstrated in studies of Qigong masters involved in mental and physical standoff.[42]

Whenever we attempt to interact closely with someone else, for good or ill, the first impulse of our brains is to copy each other.

Besides internalizing experience outside ourselves, we also perceive the world through an invisible and constant conversation with our surroundings. In 1970, while investigating a cure for cancer, a German physicist named Fritz-Albert Popp stumbled on the fact that all living things, from single-cell plants to human beings, emit a tiny current of photons, or light, which he labeled "biophoton emissions." Popp immediately understood that a living organism makes use of this faint light as a means of communicating within itself and also with the outside world.

Popp and some forty other scientists around the globe have carried out more than thirty years of research on biophoton emissions. They maintain that this faint radiation, rather than DNA or biochemistry, is the true conductor of all cellular processes in the body. They have discovered that biophoton emissions reside within DNA, setting off certain frequencies within the molecules of individual cells.[43] When first taking these measurements, Popp and his colleagues used exacting equipment capable of counting light emissions, photon by photon, enabling them to discover something remarkable. When skin ointment was applied to one part of the body, a large change occurred in the number of light emissions not only where they'd applied the ointment, but also in distant parts of the body. Furthermore the size of the change correlated in every location. Popp recognized that he had uncovered the primary communication channel within a living organism, which uses light as a means of instantaneous, or "nonlocal," global signaling.

Popp also discovered that these light emissions act as a communications system *between* living things. In experiments with a number of organisms, including human beings, he discovered that individual living things absorb the light emitted from each other and send back wave-interference patterns, as though they are having a conversation. Once the light waves of one organism are absorbed by another organism, the first organism's light begins trading information in synchrony.[44] Living things also appeared to communicate information with their surroundings—bacteria with their nutritional medium, the inside of an egg with its shell. These "conversations" also occur between different species, although the loudest and best are reserved for members of the same species.[45]

Popp and his research team were intrigued to discover that the emissions of an organism have measurable differences during daytime and evening, and also follow weekly and monthly patterns mirroring solar activity.[46] Independently he confirmed the central thesis of Franz Halberg's work: a living thing is constantly marking time with the sun.

Popp's work demonstrates that with this tiny current of biophoton emissions we create a quantum Bond with our world. With every waking moment, we are taking in something else's light.

The work of these scientists—Graham Fleming, Randy Jirtle, Franz Halberg, Giacomo Rizzolatti, and Fritz-Albert Popp—so seemingly unrelated, together suggest a profound and heretical view of the nature of living things. They also show that the concept we have of ourselves as a thing distinct from other things is fallacious.

The closer that scientists peer into the heart of matter, the more they realize that the most essential particles in the universe don't have a distinct identity by themselves. In fact in most circumstances two or more particles are so inextricably connected to each other that they can be viewed only as a collective.

Subatomically we are constantly trading light and energy to such a degree that we are not the same from one moment to the next. We are a dynamic system, not simply because of internal changes but because we are constantly altered by the ever-changing relationship between us and things outside ourselves.

This impulse to connect is mimicked throughout nature. The physical body, the thing to which we attribute most of our individuality, is created through so many complex interactions with its environment that it cannot be considered to exist independently. Since Jirtle's groundbreaking discovery an entire field of biology—epigenetics—has been devoted to the study of how we are shaped not from within, but largely by agents outside ourselves. It has begun to dawn on scientists that biology is a Bond between outer and inner forces, formed largely from outside-in, and it is this Bond—the subtle mix of environmental influences on our biology—that we inherit. What drives evolution is not the individual gene but the Bond we make with our world. Any organism is the summation of its connection with the environment in material form.

Despite our view of ourselves as the most influential entities in the universe and at the very apex of the chain of being, the new science of chronobiology demonstrates that we and all other living things on Earth are part of a vast, complex energy system at the whim of the universe's planetary geomagnetic activity. It is our Bond with the cosmic *zeitgeber* that is largely responsible for our health, for our physical and mental stability, and possibly for much of what we consider our uniquely individual motivation.

New discoveries in neurology reveal that our constant impulse is to merge; an individual understands action outside of himself by re-creating it within, so that the observer undergoes the experience of the observed.[47] As soon as we engage with the world even the most introverted and antisocial among us creates an immediate, involuntary connection.

These discoveries are unsettling and beg some basic questions: If everything is essentially just a field of energy involved in a giant trade and is constantly shape-shifting, can we consider anything or any person a thing-in-itself? When you get right down to it, on our most fundamental level, do you or I actually have an identifiable and immutable self? If I'm constantly trading bits here and borrowing other bits there, where exactly does the world end and I begin? How can I say with any finality that this alone is *me*?

Andy Gardner, a biologist at Oxford University, has been examining whether any society is so advanced and smoothly functioning that it can qualify as a *superorganism,* a single organism in its own right. Thus far he has managed to locate two model types of community: ants and bees. These animals are such team players that all conflict has been eliminated. Individual bees and ants continuously act selflessly and are willing to sacrifice, and even die, if necessary, to protect the colony. In that sense the entire community is united in a common purpose.

A superorganism with a social organization as advanced as that of bees or ants is quite rare, Gardner believes, and can exist only when the internal conflict within a social group has been virtually eliminated. That is why, he notes dryly, "we cannot use this term, for example, to describe human societies."[48]

Nevertheless, whether or not we care to admit it, humans beings, like all living things, are part of a vast, intergalactic superorganism. Everything, from subatomic particles to single-cell organisms to the most distant stars in the galaxy, all are part of an indivisible Bond.

Even our social behavior is more ant-like than we'd care to admit. Despite our propensity for one-upmanship and competition, our most basic urge always is to connect. Human beings, like ants, are born desperate to play as a team.

PART II

THE PULL
TO WHOLENESS

Before I built a wall I'd ask to know
What I was walling in or walling out.
Robert Frost, "Mending Wall"

CHAPTER 5

BORN TO BELONG

The architect Howard Roark, played by a flinty Gary Cooper in the movie version of Ayn Rand's book *The Fountainhead,* is one of our greatest modern heroes—the very model of the resolute lone wolf outsider. He dynamites his housing project rather than allow his vision to be bastardized by "the second-handers." At his courtroom trial, armed only with his uncompromising words, he defends himself against charges of vandalism by delivering a paean to the individual.

"The creator who lives for his work needs no other men," Roark announces, wheeling around to face the jury. "I do not recognize anyone's right to one minute of my life, nor to any part of my energy. . . . No matter who makes the claim, how large their number or how great their need . . . the integrity of a man's creative work is of greater importance than any charitable endeavor."[1]

At the end of the book Roark triumphs, inexplicably managing to avoid a prison sentence despite having blown up a city block. He gets to build his building his way—and he gets the girl. In the final scene he stands astride a platform on the top of his skyscraper, his silhouette towering above the New York skyline, a colossus, the epitome of Nietzsche's Übermensch—the perfect candidate for a heart attack.

Although Rand championed the supremacy of the individual through Roark, who served as the mouthpiece for her Objectivist philosophy, the libertarian ideal—living for the self above all else—is in fact a highly toxic state of being. The idea that any of us can live life successfully with fist raised, *contra mundi,* is a fiction. Our biology is the product of fusing with the world around us. The Bond we make with the world is automatic and intrinsic to us. We are who we are only because of our connection with everything outside of ourselves.

But a similar Bond governs our social behavior. We also create a super-organism in our relationships. Whenever we interact with another person or group of people we display a deep and automatic impulse to connect with them, and we make constant demonstration of this impulse in our actions, emotions, behaviors, and opinions. To satisfy our deepest need, which is to merge with others, we constantly seek synchrony.

Most of us are taught that the single most important impulse we have is to survive at all costs. The German philosopher Friedrich Nietzsche proposed that the driving force of all human motivation is a "will to power," which he considered even more basic than the will to survive.[2] Nietzsche believed that the will to power was the fundamental force holding everything in the universe together.

Yet as the latest science demonstrates, our urge for relationship is more fundamental still. Rather than a will to power, the essential impulse of all life is *a will to connect*. Our natural instinct is to merge with the other, to move away from the atomization of our individuality to the holism of the group. Deep connection rather than competition is the quality most essential to human nature; we were never meant to live a life of isolation and self-serving survival. Human beings need part-nership just to survive; we experience the greatest stress and the most serious illnesses when we are isolated from others and from a sense of connection.

This impulse to seek connection has four signatures in any society: a need to *belong,* a need to *agree,* a need to *give,* and a need to *take turns*. Hardwired in our basic biology, these impulses get revealed in the connections we make with our nearest and dearest as well as with every person with whom we come into contact.

The most fundamental of these needs is a sense of belonging. Humanity is profoundly tribal; we feel most at home in small clusters in which we are a part of the whole. Indeed so primal is the need to belong that ostracism is one of the most unbearable situations human beings endure. Robert L. Bear, a former Mennonite, referred to the Amish practice of "shunning" as a "living hell of torture."[3] Teenage girls at

British boarding schools exile friends who have become too arrogant "to Coventry" (no one speaks to them for a term) as the ultimate punishment to get them back in line. Aborigines reserved the immense life-or-death power of ostracism for extreme cases, as it often proved fatal. This most primal of human urges—not to stand apart but simply to fit in, particularly with the people who immediately surround us— may well be so necessary to our existence that not satisfying it can be a matter of life or death.

In the final years of the nineteenth century the French social scientist Émile Durkheim was struck by a singular conundrum: why certain social groups have higher suicide rates than others. Durkheim, a Jew and one of the first academics to use a scientific approach to the study of society, had been drawn to the subject from an interest in whether societies could remain coherent and whole in the face of creeping ethnic diversity and the wholesale decline of religion. It was Durkheim who first coined the term "social integration," for even in those fledgling days of sociology he understood that the fabric of society could exert a powerful force on the individual. He was particularly interested in conditions in which the society and the individual were at odds with each other, and how this might predispose a person to kill himself, and he drew on this fascination in what became a classic study of suicide.[4]

Even today, among many scientists, suicide is largely considered a private affair; the individual's reasons for doing away with himself are assumed to have nothing to do with the society in which he lives, and indeed most studies of suicide even after Durkheim's findings have focused on individual motivation. In his own work Durkheim discovered that suicide rates were higher among people without strong attachments—the childless, single, widowed, or divorced—compared to those who were married or had children, which is hardly surprising, as those with loved ones are presumed to have someone to live for.

What was surprising, however, was the large disparity in the suicide statistics of Catholics and Jews as compared with those of Protestants. The stark fact was that Protestants killed themselves far more often than Catholics and Jews did. Part of the difference could be explained away by the strong prohibition against suicide in the Catholic faith, which considers it a sin so unforgivable that it damns the perpetrator forever.

Nevertheless this factor had to be partially discounted, as the same low suicide rate was present among Jews. Durkheim concluded that it must have to do with stronger social attachment and control in Catholic and Jewish society, since Catholic countries and Jewish communities, in the main, maintain stronger family and societal ties.

There were discrepancies in his argument; Quebec, Canada, a highly Catholic province, had an extraordinarily high suicide rate, particularly among young people, but this, as Durkheim saw it, had to do with the breakdown of communal religion—again, an essential scheme in the societal fabric.

Eventually he understood suicide to be an extreme response to the severing of the social contract; as human beings need to be strongly attached to their group, so those choosing to commit suicide do so because they have, for some reason, failed to be integrated into their society. Suicides suffer from, as Durkheim put it, "excessive individuation." In other words, people kill themselves because they feel left out, and not fitting in is the one thing they cannot bear. The suicide essentially thinks before jumping, *Goodbye cruel world.*

Even with this crude first research Durkheim recognized that the solution to individual suicide lay in fixing the individual's relation to society, not the individual himself. Durkheim's prescience was validated in 2009, when Mary Daly and Daniel Wilson of the Federal Reserve Bank of San Francisco, teaming up with Norman Johnson of the Census Bureau, examined suicide deaths in the United States to see if they had anything to do with income. Their initial assumption was that they'd find clusters of suicides among the very bottom rung of the income bands of any given part of America.

At first glance it appeared that the less income someone made, the more likely he was to kill himself. Closer analysis, however, proved this conclusion to be premature. It was true that individuals in the very lowest income bracket, with family income below $20,000 in 1990 (the equivalent of about $31,000 in 2006), were significantly more likely to commit suicide than those with income above $60,000. But for anyone making anything more than $20,000, income had no significant effect on suicide risk. The only time individual income mattered in any way was in comparison to the level of income in a person's own county.

As it happened, the richest areas of America had the greatest risk of suicide.

Daly, Wilson, and Johnson also examined whether this association had anything to do with the high cost of living, the high cost of housing, the difference between renters and owners, the cost of living across the entire state, the reporting biases about suicides, or even low access to emergency health care. None of those factors appeared to have any bearing on suicide risk. The one factor that ultimately proved to have a connection with the desire to commit suicide was comparing your own income to that of others around you. And the richer everyone was around you, the more miserable you were likely to be.[5]

In the crudest terms, for every 10 percent more your neighbors make than you do, your suicide probability increases by 7.5 percent.

Simple desire to keep up with the Joneses, the research team concluded, was the most likely reason for self-harm. And the higher the benchmark, the more affluent the neighborhood—in short, the higher the stakes—the more likely the inhabitants were to find themselves falling short and the more likely they were to kill themselves.

Income envy is also rife in Europe. The economists Andrew Clark and Claudia Senik from the Paris School of Economics showed that Europeans also constantly judge themselves according to where they stand on the economic ladder in comparison to everyone around them. In the European Social Survey, which polled 34,000 people from twenty-three countries, Clark and Senik found that three-quarters of the participants believed that it was important to compare their own earnings against others.[6]

Nevertheless the more they did so, the unhappier they became.

The researchers judged levels of happiness according to their participants' responses to questions asking whether they believed that they lived comfortably, felt optimistic, had been depressed recently, and felt satisfied with their lives thus far. Clark and Senik made another remarkable discovery: a ranking existed in the participants' minds about the various categories of people they were being compared with and which comparisons cut the deepest.

The least harmful to overall happiness were comparisons made with coworkers, while income envy of family members proved to be far more corrosive. Most toxic of all, however, were comparisons with one's closest friends, which were considered twice as damaging as the more neutral comparison with work colleagues.

The most fascinating aspect of the Daly study and the European survey is that they scientifically validate the truism that money—even a collective improvement in finances—doesn't buy happiness. An overall increase in standard of living apparently has nothing to do with creating a state of contentment. Conversely, self-declared happiness also had no correlation with any objective benchmark, such as increase in an area's per capita income, or even personal benchmarks, your expectations of what you *should* be making—only the benchmarks set by your community. This is referred to as the Easterlin Paradox in psychological studies; it means that there is no objective measure of success, only individual measures formed by our aspirations and expectations and those of our nearest and dearest. With success in modern Western society defined by the showy accomplishments of individuals, a yardstick—in the form of your neighbor's goods—is always on hand to measure his pile, his status, and even his children against yours.

Our deepest need, this would suggest, is to belong to a community and in modern-day terms this translates into avoiding becoming a financial outcast at all costs. So powerful is our longing for connection that the quality of this Bond we make with a group is the single most important key to health.

In 1955 Len Syme, a young sociology fellow at Yale, made what was then considered a naïve and reckless decision, something virtually unheard of in either sociology or medicine: to study the sociology of health. Syme suspected that social factors had a bearing on certain diseases, such as heart disease, cancer, and arthritis, which at the time were believed to be primarily due to an individual's lifestyle, with a dietary or environmental genesis. His thesis advisors urged him to reconsider; there was no literature on the subject in either sociology or medicine, and likely never to be any. Nevertheless, with a stubborn doggedness and a quiet determination to buck authority, Syme held his ground, eventually becoming the first sociologist to land a job in the U.S. Department of Health, Education, and Welfare, an appointment considered so foreign that his superiors had no idea what title to offer him.

After a first and largely fruitless attempt to tease out why rates of cardiovascular disease varied in numerous states, Syme abandoned the statistical tools he'd learned in graduate school and started to carry out what is disparagingly called in science a "fishing expedition," combing through the data to find a hypothesis—in this instance, looking for any social conditions that might vary between those who'd had a heart attack and those who didn't.

What Syme found was that the "culturally mobile," as he began to call them—those who had moved geographically from one societal culture to something quite different, particularly from farms to white-collar city jobs—went on to develop heart disease. This connection prevailed even when he eliminated other factors, such as smoking, high blood pressure, and all the other supposed major risk factors of cardiovascular disease.[7]

Social mobility—moving outside your own cluster and no longer belonging—made you ill.

Syme presented his findings to a roomful of the world's most eminent cardiovascular epidemiologists, who rejected them out of hand. After a stint at the National Institutes of Health, where he established the first epidemiological review group to fund this type of research, Syme landed a professorship of epidemiology at the School of Public Health at Berkeley, the first sociologist to hold such a position. There he teamed up with Reuell Stallones, another of Berkeley's professors, to test his migration hypothesis with the perfect population: the Japanese who had migrated to Hawaii and California.

As a people, the Japanese fascinate students of epidemiology because they are such an apparent paradox: they have the lowest heart disease in the world, despite the fact that smoking, one of the strongest risk factors, is virtually universal among men. Japan's longevity statistics confound all our expectations about what is required to live a long and healthy life. The country produces the world's largest number of centenarians; currently a reported 40,000 people have reached their hundreth birthday, many of them smokers.

Epidemiologists find transplanted societies particularly instructive, as they afford an opportunity to examine just how a particular community fares when confronted with profound social, cultural, or dietary upheaval. Syme and Stallones examined heart disease risk, plus dietary factors, and any social change in a pool of 12,000 men, divided among those who remained in Japan and two groups that had emigrated to Hawaii or northern California.

Stallones was interested in whether the Japanese had low rates of heart disease in their home country because of their low-fat diet, and whether the rate went up when they adopted a typical burger-and-fries American diet. But Syme was fascinated by the social factor: whether moving countries and cultures was so destabilizing that it caused heart disease.

The results confounded both their expectations. The rate of heart disease among Japanese men immigrating to California was five times greater than the rate among those who stayed in Japan, and the rate among those who immigrated to Hawaii was midway between the two. This signaled that immigration didn't automatically cause heart disease. Yet the results appeared to be completely independent of any of the usual supposed risk factors of heart disease, such as smoking, high blood pressure, diet, or cholesterol count; in fact the Japanese population studied contained the highest number of smokers and the lowest levels of heart disease.

Amazingly their results also appeared independent of any dietary changes. Whatever the Japanese ate—whether tofu and sushi or a Big Mac—had no bearing on their propensity to heart disease.

To try to make sense of these results, Syme enlisted Michael Marmot, one of his graduate students, to pore through the data. Marmot discovered, and ultimately wrote his doctoral dissertation on, a finding that confirmed and expanded Syme's hypothesis: although changes in the dietary habits made no difference in terms of heart disease, the kind of society the transplants created for themselves did. Syme and Marmot classified the transplanted population according to the degree to which they retained their traditional Japanese culture, including social ties. Those Japanese men who had adopted an American lifestyle suffered an increase in heart disease, while those who had retained their traditional culture had the lowest levels of heart disease.

The most traditional group of Japanese Americans had a heart attack rate as low as their fellow Japanese back home, while those who had adopted the Western go-ahead lifestyle increased their heart-attack incidence by three to five times. These differences could not be accounted for by any of the usual risk factors, such as diet.[8] Those with social networks and social support were protected against heart disease regardless of whether they smoked or suffered from high blood pressure.[9]

Symes was so intrigued by these results that he decided to travel to Japan in search of the missing X-factor of impregnable health. He interviewed scores of Japanese to find the single feature that most distinguished the social fabric of America from that of Japan. And what he found, as his interviewees repeated in interview after interview, was that Americans were lonely. *Anyone could see that. Americans even walk on the street alone.* The Japanese, particularly in southern Japan, maintain tight-knit social groups that are mutually supportive, even in business. Until Japan's severe economic recession in the 1990s, joining a firm as an employee was not unlike marrying into a family—a relationship for life.[10]

Syme returned to California and enlisted another graduate student of his, Lisa Berkman, to examine the importance of social networks and social support as protection against heart disease. Berkman assembled the health statistics of most of the inhabitants in Alameda County by laboriously combing through nine years' worth of Human Population Laboratory statistics. Eventually she was able to show that those who felt lonely and socially isolated were two to three times more likely to die from heart disease and other causes than those who felt connected to others. These results were independent of risk factors such as high cholesterol level, high blood pressure, smoking, and family history.[11] Berkman was fascinated to learn that our biological responses to stress, the "fight-or-flight" mechanisms of our autonomic nervous and endocrine systems, are subdued when a companion is present, or we believe that support will be present, or even when we just think about support. Even connection with a pet is protective. Old people with pets have lower blood pressure than those without them.[12]

Syme's early work spawned many long-term population studies, including the classic study on the effect of solidarity on health, conducted on residents of Roseto, Pennsylvania. Stewart Wolf heard by chance from a fellow doctor that the inhabitants of Roseto had virtually no heart disease. This prompted him and his colleague John G. Bruhn to carry out a thirty-year study, comparing the social and dietary conditions of the town's population with those of its neighbors.

As Wolf and Bruhn discovered, the inhabitants of Roseto had half the heart-attack rate of neighboring towns, but at first glance there was no sound medical reason for their good health. The Rosetans were a beleaguered group of some two thousand people transplanted from a single Italian town of the same name. When they'd first immigrated to America, few spoke English. For years they didn't even have a proper Catholic church in which to worship. As workers they had been discriminated against by the established Welsh American citizens and forced to do difficult and dangerous work in quarry mines while being paid far less than non-Italians. In terms of the standard risk factors for heart disease, the Rosetans should have been dropping like flies. Many of the men smoked, and most were overweight, as their meals were usually cooked in lard.

Wolf realized that this tiny town was an extraordinary crucible, whose residents had a highly cohesive sense of cultural community almost unique in America. With the arrival of a British priest to the area, Father Pasquale de Nisco, the town had flourished. Under his constant encouragement, the Rosetans had festooned the town with flowers, established the Roseto band, and created a thriving group of social organizations, from religious groups for all ages to countless secular clubs.

One generation later the cohesiveness of the town broke up; the youth didn't carry on the sense of community, and before long it began to resemble a typical American town: a collection of isolated individuals keeping up with the Joneses. By the 1960s the first Rosetans were on welfare. As these social changes became established, the rate of heart attacks quickly escalated to that of the national average. As one of the first welfare recipients complained to Wolf, "You don't understand, Doctor; things have changed. People don't care."[13]

Wolf and Bruhn found a similar situation when they examined a study comparing heart attack statistics in Nevada with those of Utah. They are neighboring states, their ethnic mix is similar, they both have similarly high education statistics. Nevertheless, these states' statistics on their rates of mortality from heart attack are on opposite ends of the spectrum. Nevada has one of the highest death rates in the country, Utah one of the lowest. Initially, this made no sense; to all appearances, Nevada should have been the state with the healthier residents because it was the more affluent, boasting a 15 to 20 percent higher average household income than Utah. When Wolf and Bruhn looked more closely, they discovered that the biggest difference between the two states lay in the stability of the social structure—the close-knit nature of the family unit in predominantly Mormon Utah, compared with the high degree of broken and dysfunctional family life in Nevada. The weakening of the social fabric, they concluded, had the most to do with Nevada's higher mortality rate.[14]

Syme concluded that the quality of the connection to one's immediate geographical cluster is one of the most potent predictors of health and illness. Even in the face of devastating adversity, a cohesive group Bond can be protective against virtually any risk factor: migration, dislocation, poverty, poor diet, even alcoholism. Studies of indigenous cultures also demonstrate that a strong social tie acts as a buffer against other supposed high-risk factors, such as a foreign diet or even a foreign religious practice. For instance, a group of researchers studying the native populations of the Solomon Islands found that they had no coronary heart disease or high blood pressure even after they had adopted a Western diet and religious practices. This puzzled the researchers until they discovered one factor that had remained constant: the social ties and roles within the family.[15]

Paul Whelton and a team of researchers at Johns Hopkins University found a similar phenomenon when they studied the Chinese indigenous people, the Yi. The slender, farming Yi, who consume a diet mainly of rice, whole grains, and vegetables, have very little heart disease and low cholesterol levels, compared to the Han, an urban tribe. Yet when the Yi migrate to urban areas they suffer a significant rise in the rate of high blood pressure that begins to mimic that of the Han.

The most interesting finding about the change from rural to urban life concerns the impact of alcohol consumption. The Yi like their drink, and in their rural setting alcohol consumption does not have significant impact on their health. But once they move to the city and feel disconnected from social support, cases of hypertension, attributed to alcohol use, sharply rise. Clearly, although the Yi maintain the same diet after they move to the city, changes in lifestyle and dislocation from their rural social community have a devastating effect on their health.[16]

As the most forward-thinking medical minds realize, the genesis of most disease is stress—not the stress of transient life events, such as the state of our finances or relationships, but the stress generated by our global response to life, how we perceive our place in the world, particularly within our immediate environment. All this research suggests that our need to belong is so intrinsic that the quality of our social Bond is fundamental to our survival. An enormous body of research reveals that the root of stress and ultimately illness is a sense of isolation, and most toxic of all appears to be our current tendency to pit ourselves against each other.

The every-man-for-himself attitude fostered in the West, particularly in American society, can prove deadly to us, particularly to our hearts. Numerous studies show that people who are Roark-like—self-absorbed, cynical, and hostile to the world—are more likely to die from a heart attack. The heart expert Dean Ornish has discovered an extraordinary statistic: all the usual risk factors for heart disease—smoking, obesity, a sedentary lifestyle, and a high-fat diet—account for only half of all heart disease. Every so-called lifestyle risk factor laid at the door of cardiovascular illness by the medical community has less to do with someone having a heart attack than does simple isolation—from other people, from our own feelings and from a higher power.[17] In that sense, heart disease can be viewed chiefly as a disease of emotional alienation. Healthy adults with good support networks have lower blood cholesterol levels and higher levels of immune function than those without emotional support.[18]

A recent study by researchers at Columbia University of 655 stroke patients found that those patients who were socially isolated were twice as likely to have another stroke within five years compared with those who had strong social relationships. Isolation was the greatest risk factor, more so than having coronary-artery disease or being physically inactive. In fact the health risk of social isolation was comparable to that of being a smoker, having high blood pressure, or being vastly overweight.[19]

A group of researchers at Brigham Young University were so intrigued by these statistics that they pooled and analyzed data from 148 studies comparing human interaction with health outcomes over an average of seven years. Their stark conclusion: *relationships of any sort, good or bad, improve your odds of survival by 50 percent.* Isolation was equivalent to smoking fifteen cigarettes a day or being an alcoholic, and was twice as harmful as obesity.[20] And the level of survival advantage may be an underestimation of the benefit of healthy relationships on the participants. "The data simply show whether they were integrated in a social network," said Julianne Holt-Lunstad, the chief author of the study. "That means the effects of negative relationships are lumped in there with the positive ones."

Strong individuation and preoccupation with the self in the manner of Howard Roark is extremely bad for your health. There is even research showing that the more you use *I* words (*I, me,* and *mine*) in ordinary conversation, the more you multiply your risk of dying from heart disease. In one study, among those who had suffered heart attacks, the frequency of their references to themselves was the strongest predictor of morality—even more than blood pressure or blood cholesterol levels. [21]

The social Bond protects us, even in hard times. A sampling of Americans in the lowest income bracket suffered from virtually no depression about their financial circumstances, so long as they had the support of their congregation through regular church attendance.[22] Even when engaged in a daily struggle to survive, they were able to manage as long as they didn't do so alone. Other research shows that when men lose their jobs after company closures, one of the greatest factors of their ability to handle the stress of unemployment is having close connections. In fact close family structure and strong community support during childhood have been found to offer lifelong protection against future heart disease and other illnesses.[23]

Social psychologists at Britain's University of Exeter have shown that membership in social groups of every variety is extraordinarily strengthening—one of nature's best medicines. Their groundbreaking research shows that the most important predictor of health—even more than diet and exercise—is the number of groups to which you belong, particularly if you have strong relationships within them.[24] The higher your group membership in voluntary social organizations such as religious groups or unions, the lower your risk of death from all causes.[25]

Even catching infections appears to have far less to do with exposure to germs than with the state of your social life. Being socially isolated makes you more susceptible to infections, large and small. In one study carried out by the Psychology Department at Carnegie Mellon University in Pittsburgh, people who had the widest and most diverse number of social roles remained far more robustly immune to the common cold, while the least sociable people were twice as likely to come down with colds.[26]

"As a rough rule of thumb," wrote the Harvard political scientist Robert D. Putnam in *Bowling Alone,* "if you belong to no groups but decide to join one, you cut your risk of dying over the next year in half."[27]

Reality television shows like *Pop Idol* and *The X Factor* feed on our desire to be part of a social network. "The community feeling is part of the appeal of this kind of show," said Beverley Skeggs, a professor of sociology at Goldsmiths', University of London. "People connect to other people who have the same values, who behave in the same way."[28] We crave connection so much that we are willing to have it, if only with people up on the small screen.

Our sense of belonging is so powerful that it can even buffer us against discrimination. One of the most famous studies of the individual's relation to the dynamics of groups in an authoritarian environment is the Stanford prison experiment, carried out in 1971. Philip Zimbardo, a psychology professor at Stanford, created a mock prison in which a group of middle-class young men, especially screened to be the most psychologically stable of all those who volunteered, were randomly assigned the roles of guards and prisoners, with Zimbardo assuming the role of prison supervisor.

The experiment quickly got out of hand. The "prisoners" were given prison uniforms, referred to only by number, and made to follow arbitrary orders with equally arbitrary punishments in order to mimic the dehumanizing aspects of prison. The "guards" became more and more demanding and aggressive in creating and enforcing rules, finally subjecting the prisoners to degrading and even pornographic tasks, and the prisoners accepted their humiliating treatment, even though they knew they could leave at any time during the experiment. Zimbardo himself had gotten so drawn into his role of prison supervisor that it took an outsider—a student visiting the experiment—to alert him to how badly the situation had deteriorated, after which he called a premature halt to the experiment. The truthsayer eventually became Zimbardo's wife.[29]

This pivotal study has been cited in psychology classes ever since as proof positive that groups have an automatic *Lord of the Flies* effect, causing people to shed their moral judgment, even their humanity. It's also been seen as an example of how members of a disempowered group quickly lose their sense of individual identity.

In 2002 two British social psychologists, Alexander Haslam at Exeter and Stephen Reicher at the University of St. Andrews in Scotland, wished to revisit these ideas by re-creating the Stanford University experiment. Funded by the British Broadcasting Company (and broadcast on TV as *The Experiment*), the two psychologists created an elaborate prison and, like Zimbardo, randomly assigned a group of men to the roles of either prisoners or guards.

Over eight days the experimenters witnessed stunning developments. Although the prisoners were initially demoralized, the group dynamics shifted over time. As the prisoners began to develop a sense of shared identity, they began to function effectively as a group and also enjoyed improved morale and mental well-being. This shared identity also led to improved health profiles; the prisoners experienced a lowering of cortisol, the hormone pumped by the body in times of stress.

As the experiment carried on, the prisoners became stronger, happier, and more empowered. The guards, on the other hand, who had not bonded as a group, became increasingly dispirited and powerless, exhibiting high levels of cortisol. Eventually the prisoners staged a breakout and the authority of the guards collapsed.

Even as a beleaguered group, the prisoners remained robustly immune to treatment by the guards—so long as they could connect with others in the same boat. In fact the more they were oppressed as a group, the stronger they became.

The need to move beyond the boundaries of ourselves as individuals and to bond with a group is so primordial and necessary to human beings that it remains the key determinant of whether we remain healthy or get ill, even whether we live or die. It is more vital to us than any diet or exercise program; it protects us against the worst toxins and the greatest adversity. The Bond we make with a group is the most fundamental need we have because it generates our most authentic state of being.

We also experience our need to bond on an individual level in our relationships. In every way, with every part of our body, our need to connect manifests in another way: a need to agree.

CHAPTER 6

BORN TO AGREE

In the early evening of September 23, 1954, Constable Alex Deeprose of the Glasgow Police Department was called to a disturbance at Southern Necropolis, Glasgow's hundred-year-old, twenty-one-acre cemetery, only to discover hundreds of children under fourteen, many as young as four, armed with kitchen knives and an assortment of sharpened sticks, combing the 250,000-grave cemetery, fueled with purpose. They were there, they told the constable, to hunt down a vampire who'd kidnapped and eaten two of the local boys. Many of their parents not only had sanctioned the vigilantes, but were also out in force themselves.

No one could recall where the story came from or how it started, but within a few hours it had blanketed three local primary schools. By 3 o'clock, when school let out, virtually the entire student population of the immediate Gorbals and Hutchesontown areas of the city had bolted toward the giant gothic arched entrance of the Necropolis. For several hours the children had crept along together over crumbling headstones, taking note of any sinister figures, backlit by the smoke and red light that flared from the steelworks adjacent to the graveyard. Edward Cusick, headmaster of St. Bonaventure's Primary School, came forward that evening to assure the anxious parents that there were no records of any missing children in Glasgow at the time. Nevertheless night after night the vigilantes returned after sunset to continue their search.

Alarmed teachers, newspaper editors, and some of the resident Presbyterians, casting around for the source of the urban myth, were ready to finger current standards in films, despite the fact that the local cinema had in fact never shown a vampire movie. Eventually the true culprit was identified: the American comic book. Scotland's National Union of Teachers was compelled to organize an exhibition of especially lurid examples of the horror end of the genre, *Tales from the Crypt* and *The Vault of Horror,* as proof they were to blame for inflaming the imaginations of Scottish youth. Shortly thereafter the British government passed the Children and Young Persons (Harmful Publications) Bill, prohibiting the importation of magazines or comic books portraying "incidents of a repulsive or horrible nature" to minors, a law remaining on the books to this day but never actually enforced.

Tam Smith, who was eight at the time, would later recall that most of the children from the cramped wards of Gorbals did not own a television set and had neither the cash nor the wherewithal to buy these exotic publications.[1] In fact most Glaswegian children had no idea what a vampire actually was. The description offered to Constable Deeprose was of a creature seven feet tall and possessing a set of giant iron teeth. The only monster Tam had ever heard so described had been read out to him from Daniel 7:7 in the Bible. A monster of the same description had also featured in a poem read to him by one of his teachers.

The Gorbals Vampire incident is one of numerous examples of *psychogenic illness,* or mass hysteria, a word coined by Hippocrates in 400 BC to describe an inexplicable mass contagion of emotion or behavior. In one of the first recorded cases Frau Troffea stepped out into a narrow side street in Strasbourg, France, on a hot day in July 1518 and began to dance. A week later a hundred people had joined her, and within a month the crowd of dancers had swollen to four hundred. The alarmed Alsatian authorities, convinced that like cures like, hired guildhalls and musicians to play pipes and drums and recruited professional dancers in an attempt to wear out the town dancers. By the end of the summer, after many dozens of dancers had died of heart attack, stroke, and exhaustion, the authorities bundled all the survivors into wagons and packed them off to a healing shrine.

Mass contagion was behind the Salem witchcraft trials in 1692, when nine-year-old Betty Parris and her cousin Abigail Williams first complained of inexplicable fits, a condition that swept through the young girls of Salem, Massachusetts, resulting in the hanging and murder of nineteen residents on charges of witchcraft. Psychogenic illness was also considered the cause of an incident referred to by two psychiatrists as "the June Bug," in which fifty-nine workers in an American dressmaking plant who claimed they were being attacked by a swarm of insects showed very real evidence of rashes, nausea, dizziness, and fainting.[2] More recently, in May 2006, some three hundred adolescents at fourteen Portuguese schools complained of rashes, dizziness, and difficulty catching their breath, forcing some schools to close. No evidence of a genuine virus was ever isolated, but all of the very same symptoms had been experienced by one of the characters in a recent episode of *Morangos com Acucar* (Sweet Strawberries), the highly popular Portuguese teen equivalent of *90210*.[3] These days social contagion can even be caught from a soap opera.

No matter what the incident, the symptoms of psychogenic illness are the same: a behavior or physical symptom that inexplicably spreads from person to person like a virus. Psychiatrists consider these incidents irrefutably mental in origin and officially characterize them as mass "hysterical neurosis, conversion type," said to be generated by the social concerns of the day. Nevertheless mass hysteria is simply an extreme version of another face of our social Bond: the need to seek accord. With our natural impulse to merge with others we have a deep urge to be the same as them, mentally and physically.

Our relationships represent a move away from the atomization of the self and toward connection with the whole. One of our most primitive urges, in every social relationship, whether individual or collective, is to agree with each other. No matter how obstreperous our nature, we constantly seek physical and psychic equilibrium with everyone with whom we are in contact, even if that means copying someone's desire to hunt down vampires. This need manifests in an automatic impulse *to synchronize,* physically, psychologically, and emotionally—to place ourselves in the same state as that of whomever we encounter. Our deep need to agree does not rest on any moral stance; the rapid rise of Nazism in Germany is an example of the contagion of a belief system and the willingness of a good percentage of the population to suspend universal human values in order to conform to it. In his book *Talking to the Enemy* the anthropologist Scott Atran convincingly argues that suicide bombers kill themselves out of a deep-seated need to be accepted by their group, not for religious reasons per se. They take on the "commitment cost" of acceptance, which depends on the enormity of the sacrifice they are prepared to make.[4] "People don't simply kill and die for a cause," writes Atran. "They kill and die for each other."[5]

According to research carried out by the psychologists Elaine Hatfield and her colleagues at the University of Hawaii, when people interact they undergo automatic and involuntary mimicry and synchrony involving their entire body. The listener quickly begins to adopt the facial expressions and posture of the person speaking. He even copies her rate of speech, the length of her utterances, and her pauses for breath. In fact for the relationship to work well, Hatfield and her colleagues believe, even their voices must become entrained.[6] In every social encounter we become an expert at mimicry for a larger purpose: a seeking, always within us, to Bond.

William Condon, a psychologist at Boston College, spent more than three decades patiently watching videos in slow motion, frame by frame, in an attempt to understand what happens to people's bodies when they speak to each other. As Condon discovered, our every movement is locked precisely with our speech patterns; our hands, arms, shoulders, and head all keep time to the beat of our speech, and even our eyes blink in synchrony. Those movements can change in a few seconds, with every new consonant and vowel we emit, and are entirely involuntary. "You can't break out of this, no matter what," Condon notes.[7]

But most surprising about Condon's videos is the reaction of the listener to a conversation: his body begins to synchronize with the speaker's speech patterns after only the tiniest of delays (forty-two milliseconds, the equivalent of one frame of film traveling at twenty-four frames per second).[8] Finger, arm, eye, and head movements all coordinate perfectly with the speaker's voice, its emphasis, pitch, and volume: the two bodies move and sway together as if choreographed in a complex dance. The big sweeping gestures of the entire arm will track an entire phrase, while the slighter gestures, down to the movement of a single finger, will punctuate a word or sound for emphasis.

"We're almost in auditory touch," says Condon. "When I speak to you, my thoughts are translated into muscle movements and then into airways that hit your ear, and your eardrum starts to oscillate in absolute synchrony with my voice. In essence there's no vacuum between us—it takes only a few milliseconds for a sound to register in the brain stem, 14 milliseconds for it to reach the left hemisphere."

Condon attributes this impulse to an auditory-motor reflex in the central nervous system, which "might allow, even *force,* a listener's movements to synchronize with a speaker's voice far faster than any conscious reaction time" (italics mine). It's even possible, he says, that we are *anticipating* speech precognitively, before the speaker even opens his mouth.

The listener's ability to body-map a speaker is inborn; twenty minutes after birth, newborns possess the ability of an adult to track the sounds of speech.[9] This occurs with any sort of speech, even a range of languages foreign to the infant's parents. Only later in their development do babies habituate themselves to the sound of their own language.

In an attempt to discover the source of this invisible connection Condon wired two people to EEGs to monitor their brain waves while they were deep in conversation. At the time the equipment used was similar to old-style lie detectors, whose recording pens tracked brain wave activity onto a roll of paper. Time after time, Condon discovered, as the pair spoke to each other the recording pens of the two machines moved in synchrony, as if emanating from a single brain. The only thing that interrupted this perfectly orchestrated duet was the arrival of a third person, at which point the listeners would begin a new tempo, synchronizing with the interloper.[10] Eventually, Condon realized that brain structures at multiple levels of the nervous system mediate this kind of microsynchrony. Quite simply, one nervous system begins to drive the other.[11]

After years of studying his films in slow motion Condon concluded that human beings are not "isolated entities sending discrete messages" back and forth, but part of a shared organizational form, as expressed through language and movement.[12] Edward T. Hall, the noted American anthropologist, put it another way: "Basically, people in interactions move together in a kind of dance . . . without music or conscious orchestration."[13]

Hall, who carried out similar studies himself, discovered deep synchrony in groups and encouraged his college students to investigate this by filming groups in action. In one instance a student of his involved in a seminar project hid in an abandoned car, photographing a group of children dancing and skipping in a school playground. At first their play appeared completely random, each child playing alone, though one young girl covered far more of the playground than her peers. When he got back to the university the student slowed down the film and played it at different speeds, as Hall had instructed him to do. Eventually he saw the children moving in synchrony to a definite rhythm. "The most active child, the one who moved about most, was the director, the orchestrator of the playground rhythm!" wrote Hall.[14]

What also fascinated Hall was that the rhythm at which the children played seemed familiar. He asked a friend expert in rock music to find a tune whose beat fit the rhythm of the children's moves. When a piece of music was located and synchronized with the children's play, Hall noted that the children kept perfect time to the beat of the music for the entire four and a half minutes of their play. In fact when others viewed the film, they believed that the children had actually been listening to music on the playground. The "unconscious undercurrent of synchronized movement was what tied the group together," wrote Hall. "Without knowing it, they were all moving to a beat which they generated themselves. . . . They even had a conductor who kept the beat going continuously."[15]

Our need to copy others occurs in virtually all behavior. We now know that yawn contagion occurs across all species and even between species; primatologist Frans de Waal once attended a lecture with a film showing horses, lions, and monkeys yawning, and before long, the entire audience was yawning in synch.[16] Even laughter has now been proven to be contagious. Research from University College London has revealed that a variety of sounds, from laughter and cheering to screaming and retching, trigger responses in the premotor cortical regions of our brain, which prepares muscles in the face to configure in ways that correspond to the sound. Remarkably, the pleasant sounds produce twice as big a response in the motor cortex as the unpleasant ones, suggesting that positive sounds, such as laughter, are more contagious than negative sounds. The trigger in the brain for copying laughter is automatic; we find it difficult to control the impulse to smile or laugh when we hear laughter.[17]

Members of virtually any species engaged in a shared activity will begin to move in synch. Packhorses, which may start out walking individually, very soon adopt a pace in step with each other and walk over obstacles as if they have become a single organism, writes de Waal. He recounts the story of Isobel, a husky sled dog who had become blind but was so used to synchronizing with her pack that she was able to keep in perfect step with them as they pulled a sled.[18]

The purpose of mimicry is to pave the way for deep emotional connection. "When people automatically mimic their companions' fleeting facial, vocal, and postural expressions of emotion, they often come to feel 'a pale reflection' of their companions' actual emotion," says Hatfield. "By attending to this stream of tiny moment-to-moment reactions, people are able to 'feel' themselves into the emotional lives of others."[19] This in itself can be highly infectious.

Sigal Barsade, a young woman in California, arrived at work one morning and noticed that the general mood of the office environment, usually so edgy and stressful, had completely changed, although she couldn't quite work out what was different. There were no new employees, no change of management, and no change of scenery, but those who had barely acknowledged her before now looked up from their work and smiled. Workers who usually spent the entirety of every day glued to their computer screens took breaks to chat around the coffee machine. For the whole of that blissful week her workmates were more relaxed and sociable than she'd ever known them to be.

In the following week the collective mood had returned to normal: tense, testy, and sullen. Nothing again was outwardly different, and yet once again the whole of the office had been profoundly affected. The only variable to which Barsade could attribute this change was the return of a petulant coworker from vacation. Even though the woman didn't work closely with Barsade, so palpable was her complaining and snappish temperament that it had infected everyone who worked around her. Perhaps people are walking "mood inductors," thought Barsade, and they pass on their moods to whomever they encounter, who in turn pass it on, in an endless daisy chain.[20]

The experience profoundly affected Barsade, largely influencing her decision to pursue a doctoral degree in the behavior of organizations at Berkeley's Haas School of Business at the University of California. What had most fascinated her about the experience with her sullen coworker was how mood contagion between employees might influence relationships, decision making, even profits. After earning her doctorate she accepted a post as associate professor at Yale's School of Management, where she decided to test what she had begun to call "the ripple effect" of emotion by devising an ingenious experiment.

She randomly assigned ninety-four undergraduate students at the business school to small clusters of two to four others and asked each participant to play the part of managers on a salary committee, negotiating how best to parcel up limited sums as pay bonuses for their employees. Each manager was to act as an advocate attempting to obtain the largest sum possible for a candidate from his department who had been put forward as particularly deserving. All bonuses were entirely dependent on each member of the group coming to an agreement within a set amount of time.

Unbeknown to the students, Barsade had placed a cuckoo in the nest, a drama student who had been specially trained to act a different mood with a different energy level in each group. "Rick," who was always assigned to represent the same employee in each group, was also always asked to speak first in order to see if his mood would set the emotional tone of the meeting.

The results, which Barsade videotaped, were striking. Even though Rick made identical requests of every group on behalf of his candidate, the mood of each group differed in accordance with Rick's mood. The collective mood and Rick's role in creating it also had a significant influence on the tone and outcome of the negotiations. When Rick exuded pessimism and negativity, the group was less likely to cooperate with one another; when he was calm and happy, the group was more likely to bond and work with each other productively. The collective mood and Rick's role in creating it also had a significant influence on the outcome of the negotiations: the more cooperative a team, the higher their bonus splits.

The effect was not only insidious, but also completely unconscious. None of the participants had the slightest idea his mood was being artificially manipulated. On the questionnaires about their feelings before and after the experiment that Barsade asked participants to fill out, all attributed their own effectiveness or ineffectiveness in the group to other factors, never to the collective mood.

Barsade made another fascinating discovery she hadn't counted on: positive emotion was just as contagious as negative emotion. This particularly surprised her because earlier research suggested that negative emotion was the more infectious. There also seemed to be little difference in group mood no matter whether Rick was acting with high or low energy, except in one regard. When Rick was subtly upbeat, his mood was more socially contagious than his subtle "low-energy" bad moods. In fact, he was most persuasive in his "low-energy" positive state; the group actually gave him *more* money than he'd asked for. Rick's extraordinary effect on the collective mood of each group extended to all encounters he had with group members on campus in subsequent months. Those with whom he'd acted positively greeted him warmly; those who had been in groups in which he had been pessimistic continued to greet him with hostility or chilly silence.

Barsade concluded that both kinds of emotion, positive and negative, are highly contagious, but that positive emotions stimulate a group to be cooperative and make more positive choices in decision making, while the reverse held true for negative emotion.[21] Without upbeat collective emotion people are bad negotiators and make bad decisions.

The results of Barsade's experiment rocked the business community and her research was cited often, eventually landing her a professorship at the Wharton School of Business at the University of Pennsylvania. In this one inspired experiment she had managed to demonstrate clearly that emotions are like viruses, not only transferring from person to person in an endless and unconscious circle of contagion, but also profoundly affecting the outcome of business encounters and negotiations. A business could make more money with a preponderance of good moods, which would constantly infect its entire workforce in a virtuous circle. Major decisions might be at the mercy of any one of the participants' moods that day. Some companies began to make use of the lessons inherent in the study. Multilevel marketing organizations such as Mary Kay Cosmetics and Neways now use songs, "reward" dinners, and national meetings to instill positive emotions in their workforce.

As Barsade subsequently discovered, emotional contagion occurs readily even in casual encounters. The simple act of sharing with another person in any way creates mood equilibrium, which Barsade began to call "collective emotional knowledge" between people. "Whenever we interact with people, we're constantly exchanging mood back and forth," she says, and this exchange of moods and actions between individuals, groups, and entire organizations is "constant, subtle and continuous."[22]

Our need to connect with others is so all-pervasive that we readily adopt their moods, positive or negative. For instance, in one study three groups of participants listened to a speech read by an actor using happy, unhappy, or neutral inflections. When the participants later rated their own emotions, their moods identically matched those of the actor while he was reading the speech. Furthermore when they were asked for their opinion of the speaker, those who had listened to the unhappy tone were the least fond of him.[23]

We not only mimic others' emotions, but we also feel those emotions deep within our body. We are so attuned to the emotional landscape surrounding us that a positive or negative environment affects our body and its ability to function. Natural killer cells—the immune system's front line of defense against cancer and many viruses—are profoundly reactive to stress in our lives, particularly social stressors.[24] Large dips in the number and activity of these cells have been observed during arguments and even minor conflicts.[25]

Similarly social stress has been shown to affect the functioning of the hypothalamus-pituitary-adrenal axis, one of the chief regulators of the body's ability to fight off disease. The psychologist David Spiegel and his colleagues have found a link between marital discord and negative effects on the cortisol rhythms of the body, which are now considered a risk factor for early cancer mortality.[26]

The ultimate point of any social encounter is to create a melding of yourself with the other person. Our instant and automatic mimicry of movements, figures of speech, and emotions were all designed to reinforce the Bond between us. Whether or not we wish to be, we are constantly attuned to the emotional landscape around us. This means, however, that any thought or gesture may not be wholly and exclusively our own, but derive from this connection with others. In fact we can even get infected remotely, and by people we may not even know.

In 1948, in an attempt to discover the common causes of cardiovascular disease, a group of scientists at Boston University hit upon the idea of tracking its development over time among a large group of participants— in this instance, virtually an entire town. Three generations of a good portion of the population of Framingham, Massachusetts, have now been followed for more than sixty years by the Framingham Heart Study, its participants scrutinized in every conceivable way so that virtually every aspect of their health, lifestyle, and social capital has been laid bare. Offering as it does a goldmine of information about the health and habits of an entire local population, the data of the Framingham Study have been combed through by many other researchers for clues about how lifestyle influences health and psychological patterns and what happens to people's health when they live alone or in groups.

To Nicholas Christakis, a professor of sociology and medicine at Harvard, and James Fowler, a professor of political science and medical genetics at the University of California at Davis, the Framingham population represented an extraordinary opportunity to study their joint passion: how a person is affected by his social network. In earlier work Fowler had been looking at how voting behavior spreads among people, and Christakis had discovered that the health of one spouse usually affects the health of the other, but now he wished to see if this were true not simply between partners but also between friends.

As a sociologist Christakis studies how and why "nodes"—individual human beings—cluster together, and how individuals are affected when they form these groups. Sociologists love to map these dense interrelationships and then study whether individual nodes get any further value from the people within their network.

In social-network analysis the behavior of the individual isn't primary; the sociologist is looking for the *effect of the group relationship* on the individual.[27] It is the extent or quality of the ties, rather than an individual's behavior, that most holds his interest.

A social-network diagram is a thing to behold: an intricate web of lines between individual nodes, not unlike an airline's flight map, although a large and truly comprehensive network like that of the Framingham Study might appear to resemble every flight traveling from every port in the world to every other. In these diagrams, sociologists search for *homophily,* the tendency of people to choose and relate to others who are most like them. They also find a network's *components,* little clusters or groupings in which each member has at least one connection with another person in the cluster. Scientists repeatedly witness the birds-of-a-feather aspect of a social network: similarity breeds connection. In every social connection of every shape or size—the people we marry, the friends we seek out, the memberships we join, the information, advice, or support groups we elicit—we tend to choose networks that contain people with similar demographics, behaviors, and personal characteristics. The closest connections (and also the biggest divides) occur across race and ethnic lines, after which we get united or divided according to age, religion, education, occupation, and sex, roughly in that order. Furthermore groups that are unalike fray more quickly than groups that are alike.[28]

Another aspect of sociological study is what is referred to as *geodesic distance,* or degrees of separation. This refers to how close you are to others in the group, with each layer being represented by a "degree." There is one degree of separation between you and your friends, two degrees between you and your friends' friends, and three degrees between you and the friends of your friends' friends.

Christakis and Fowler picked 5,124 second-generation Framinghamians who had been enrolled in the study since 1973, plus every person with a relationship to these individuals (including friends, siblings, spouses, and children). By the time they were finished they had mapped more than 12,000 people who were in some way inextricably interconnected.

Their first inquiry was into obesity. They had information about the body-mass measurements of their entire Framingham cohort, and also their social ties and how they evolved over time. This enabled the scientists to examine whether weight gain in one person was in any way influenced by the weight gain of that person's friends, siblings, neighbors, or spouse. The central question they were teasing out was whether fat people seek out other fat people as friends, or whether their unhealthy eating habits affect their friends.

Christakis and Fowler discovered that overweight occurred in curious clusters over thirty-two years. Obese people were more likely to have networks of fat friends across three degrees of separation (the friends of their friends' friends) and vice versa.

The clusters didn't appear to be only by choice, the effect of positive selection of social ties among other fat people. Obesity was actually *infectious:* the likelihood of your becoming obese increased by 57 percent if you had a friend who became obese in any set time period, and the greatest influence was from your friends of the same sex. Fat contagion was even stronger coming from a friend than from a family member or spouse; if your sibling became fat, it increased your likelihood of becoming fat by only 40 percent, and if your spouse gained weight, your chances of following suit were increased by only 37 percent.

Christakis and Fowler were also surprised by the lack of a geographical effect. Fatness had nothing to do with proximity, with being close to other fat people. In fact the researchers found no such effect between neighbors. They were also able to discount the effect of individual lifestyle or behavioral changes, such as quitting smoking.[29]

Christakis then decided to apply this same model to emotion, particularly happiness. Happiness has been exhaustively studied and has now been identified by the World Health Organization as a key component of health. To date the data suggest that happiness is caused by a large group of factors: income level, socioeconomic status, job satisfaction, our marriage, our health, our genes, even how we feel about the current occupant in the White House or whether we've ever won the lottery. But no study up to that point had examined whether happiness had anything to do with the happiness of others in one's social circle.

The central question, to their minds, was whether individuals are innately happy or unhappy or whether their life's circumstances, *largely set in motion by social context,* affect their happiness levels over time. If happiness followed the path of other aspects of sociology, individuals would be influenced by their location in a social network and by the overall happiness of those who are closest to them in the network.

Christakis and Fowler returned to the Framingham data to discover whether happiness is contagious and whether there were such a thing as happiness "niches" within social networks. As with obesity, they found that the relationship between happy people and their friends who were happy extended up to three degrees of separation.

Their analysis of the data shows that happiness is definitely contagious, not self-selecting; clusters of happiness resulted in part from the *natural spread of happiness,* and not just from a happy person's tendency to find other happy people to be with. People who were surrounded by many happy people and were central in the network were more likely to be happy in the future. Happiness, like obesity, was socially contagious.

Christakis and Fowler unearthed another remarkable finding: unlike the spread of obesity, in the case of happiness geography did apply. The strongest effects occurred among people close by, and the effects faded with distance and separation. A friend living within a mile of you who is happy increases the probability that you will be happy by 25 percent; a happy next-door neighbor increases your happiness probability by 34 percent.

Friends matter more to individual happiness than spouses or relations. A happy spouse increases your chances of being happy by only 8 percent, and a happy sibling by only 14 percent. Coworkers also don't have as much of an effect on your happiness as close friends and neighbors.

Christakis and his team concluded that happiness is a "collective phenomenon," and one affected largely by proximity.[30] Those people you surround yourself with, with whom you have regular physical contact, in short *the friends in your psychic neighborhood,* are the people who help to determine how happy you are.

The same phenomenon occurs with happiness's opposite: loneliness. Lonely people were more likely to spread loneliness and, ultimately, to unravel the network. Christakis, Fowler, and other researchers went on to investigate a host of behaviors and found that many were contagious: runs on the bank caused other people to rush to withdraw their money; suicides among young people caused a rash of copycat activity; cheaters on an exam begat a spate of other cheaters.[31]

The sociological evidence of the contagion of happiness and sadness suggests something quite remarkable: of all your relationships, of all the people capable of making you happiest or irritating you the most, those who have the greatest effect on your mood and even your state of health are those who are closest to hand. Just as we are infected by the individuals around us, so we are largely a creation of the networks of which we are a part. The collective personalities of these networks spill over into us, defining us, ultimately helping to determine our happiness or misery, our behavior, even the state of our health.

Our psychological landscape is a complex mix of inner and outer circumstances. Our inner states can be viewed as entirely the result of outer circumstance, and even our outer states are designed and influenced by the interaction between us and the world. This complex mix of copying back and forth forms the narrative of our personality.

All this trading off of habit and emotion tends to make things a bit messy. If my experience gets so jumbled up with yours, it grows increasingly difficult to separate out what has come from me and what has come from you. Even emotion is virtually impossible to unravel and identify as uniquely one's own. We have hitherto viewed emotion as wholly individual, but it is now understood to be a complex interplay of influences between ourselves and all the others with whom we are in contact. Nature has designed life *only* in relationship, and whenever we do not recognize this we act against our own fundamental design. What we regard as our immutable and distinctive personality is actually a relationship—our Bond with the world.

CHAPTER 7

BORN TO GIVE

Samuel Oliner's life has been haunted by a single question, beyond why he, alone of the inhabitants of his entire village, managed to survive a Nazi pogrom. For six decades he has been demanding to know to why a person would be willing to put everything on the line—including the lives of her entire family—in order to save someone she barely knew.

In the summer of 1942, when Oliner was twelve, his family, like many other Jews, were forced to leave their home in Bielanka to take up cramped quarters in a ghetto in Bobowa, a small town in southern Poland. Early one August morning the *Einsatzgruppen,* a cadre of enormous trucks, roared into the middle of the plaza. Armed soldiers, Germans and Ukrainians serving under the Nazis, flooded out and banged on the doors of all the houses, demanding entry. Heeding his stepmother, who implored him to save himself, Oliner hid beneath their sloped roof. Through a tiny aperture, he was witness to unspeakable atrocities—a infant girl casually flung out of a top floor window as if discarded, another crying baby silenced by gunshot fired by a soldier who'd just finished raping its mother—until the Nazis herded all the survivors, including Oliner's family, into the trucks and the shouting and screaming yielded to a terrible silence.

After the *Einsatzgruppen* departed, Oliner ran barefoot through the countryside, sleeping rough and managing to elude his countrymen who were rewarded for reporting stray Jews to the Gestapo. By chance, several days later he finally learned the fate of his family: they and a thousand others had been taken to the forest of Garbacz, where they were stripped, made to stand on a plank, and systematically shot by soldiers with machine guns so that they would fall into a giant mass grave that had been dug below where they stood. It had taken eighteen hours to kill that many people. Some of the wounded were simply left to suffocate under the dead, buried alive.

Oliner managed to make his way to Bystra, another village, where he knocked on the door of the home of Jacek and Balwina Piecuch, a Christian family he barely knew. Balwina had attended school with his father as a young girl. She had heard about the mass extermination at Garbacz, and as soon as she opened the door and saw Oliner she hugged him to her and ushered him in.

For the following three years Balwina made sure Oliner remained safe from the Nazis. She gave him a new name, taught him to act like a Christian, secured him a job with one of the Polish farmers, and continually offered love and reassurance when he was close to despair. Oliner survived and immigrated to America, where he married and became a noted sociologist. But over the years he asked himself constantly: *Why had Balwina done it?* Her decision to help him had put her family in extraordinary jeopardy, for her house was surrounded by countless informers who would receive generous rewards from the Nazis for reporting Jews in hiding. What made her risk everything—her own life and the lives of her husband and two children—for a relative stranger?[1]

The question burned so long and hard within him that eventually he was compelled to ask it of anyone he heard about who had behaved altruistically on that heroic scale. Oliner conducted a lifelong study of the motivation of ordinary people performing extraordinary acts, those who had rushed into burning buildings or dived into icy waters to save other lives. *Why did they do it?* What prompted them to risk their lives for another person, even a stranger? Actions like Balwina's contrasted utterly with those of her neighbors; in fact, they flew in the face of every truth he'd been taught about the essence of humanity—which is supposed to operate from a solid core of self-interest.

Every story we are told about ourselves maintains that if left to our own devices, without the taming influence of religion or a social contract, we would act according to our true natures: cold-bloodedly and entirely for self-preservation. "Let us try to teach generosity and altruism," writes Richard Dawkins, "because we are born selfish."[2]

From this perspective actions such as Balwina's appear to be errant behavior or an error in judgment. After all, acting unselfishly out of

concern for the interests of others, regardless of the personal consequences, can be deleterious to the self and even reduce the possibility
of survival. Altruism makes no logical sense because it is potentially an
act of purposeful self-destruction. In a zero-sum game it means deliberately choosing the shorter straw.

Evolutionary biologists such as Robert Trivers of Rutgers University
believe that altruism represents a neural error—the brain's mistaken
response to something untoward. "Our brains misfire when presented
with a situation to which we have not evolved a response," he notes.[3]

Cases of altruism wreak havoc with classic ideas about survival of
the fittest. Consequently, scientists have attempted to shoehorn altruism into current biological theory by reducing unselfishness in animals
and humans to a genetic imperative: acts of self-sacrifice occur only
because of genetic favoritism. Many modern biologists represent altruism as an equation, measuring its cost or benefit in terms of its bearing
on an organism's number of offspring, or "reproductive fitness." Altruism increases the reproductive fitness of another being and ultimately
costs the altruist offspring or genes.

An eccentric American scientist called George Price became obsessed
with the origins of kindness since it did not square with evolutionary
theory as it was understood at the time. In pursuit of the source of
kindness, Price rather unkindly took a permanent hiatus from his wife
and child, moving to London in the 1960s to seek out the evolutionary
biologist W. D. Hamilton. Together they formulated a mathematical
equation for altruism, using economic theory to work out how altruism
aids evolution.

Their theory of kin selection, or "inclusive fitness," explains altruistic behavior as a means of perpetuating the family line. An animal
that offers itself as a helper increases its own chances of survival and
future reproduction or ensures the survival of the family.[4] Birds feed the
young of relatives because that behavior increases the number of their
shared genes in future generations.[5]

Another variation on kin selection is group selection, or group adaptation, the theory that individual actions operate on behalf of the group's
gene pool. This is thought to occur because most of the members of the
group in question are ultimately related. Dawkins also offers a cost-

benefit equation that calculates the point at which it becomes genetically advantageous for an animal to display altruism. The gene that promotes the desire to help other family members aids its own replication, which helps the spread of individuals bearing copies of that gene.

The selfish-gene theory argues that animals, including humans, are merely "robot survival machines" shaped by the desperate survival imperative of their genes. And it is this primary impulse, propagation at all costs, that ultimately accounts for the human characteristic of egocentricity. "This gene selfishness," declares Dawkins, "will usually give rise to selfishness in individual behavior."[6]

The problem with the selfish-gene theory, and indeed all theories attempting to rationalize altruism from the point of view of survival, is the vast number of exceptions to the rule. Research of every variety offers instances in which animals behave in what could be called "random acts of kindness": extraordinary self-sacrifice, compassion, courage, and generosity toward members of their own species, members of other species, and even toward humans, often to their own detriment. During Darwin's lifetime Russian scientists rejected the aspect of Darwinian theory related to competition within species as simply characteristic of the English penchant for individualism. Biologists such as Petr Alekseevich Kropotkin formulated an alternative theory called "mutual aid," which posited that the primary struggle for existence concerns hostile elements in the environment, such as harsh temperature, but that animals join forces and cooperate with each other in that struggle. Natural selection favors cooperation, not competition, Kropotkin asserted in his book *Mutual Aid*.[7]

A basic drive for cooperation and partnership, even sacrifice, rather than selfishness and naked survival, appears to be intrinsic to the biological makeup of all living things. Animals of all persuasions carry out an enormous coterie of actions that have no raison d'être other than to help those less fortunate or maintain the cooperative social glue.

Animals often partner up with members of unrelated species; badgers and coyotes will often form pairs to hunt.[8] Animals within the same group form buddy systems, in which the more successful hunter helps the less fortunate. In the wild, vampire bats who have had a successful feeding spree on cattle routinely regurgitate the blood, which they will donate to those among their group who have been less successful.[9] Many species of animals employ alarm and information systems for danger and food, even when this puts them at risk. Vervet monkeys use alarm calls to warn other monkeys of impending attack, even though raising the alarm can increase their own chances of being harmed.[10]

Some of the most extreme instances of altruism involve animals that adopt unrelated animals.[11] There are even instances of an animal from one species adopting those from another.[12]

Animals also routinely put aside the most fundamental drive of all: the need to eat. They share food or ensure that weaker individuals in a pack or herd are fed, even when they have to give up their own food.[13]

The Dawkins's-eye view of the universe would argue that altruism is impossible among animals that are not closely related because it runs counter to survival. Yet a recent German-American field project with primatologists in Kibale National Park in Uganda, studying the social relations of an entire colony of chimpanzees, found that although chimps spend more time with their relatives, they also are highly cooperative among those with whom they lack any sort of genetic kinship. Kevin Langergraber and his colleagues concluded that in the majority of close and cooperative affiliations between male chimps, the animals were unrelated or distantly related. Genes played no part.[14]

Altruism has even been displayed by animals in laboratories. In one experiment two rats, strangers to each other, were placed in a cage, but one was suspended in air by a harness. As the airborne rat shrieked for help, the other rat showed signs of distress and appeared to want to try to help. It soon discovered that by pushing on a bar it could lower the suspended rat to the cage floor. Although its fellow rat was neither a relative nor a mate, and there certainly was no survival advantage to himself in helping the other rat, it would not rest until it had alleviated the other rat's suffering.[15] Similarly studies of monkeys have shown that they will choose to starve themselves for days on end if that will spare other monkeys from receiving a shock.[16] Even rats won't eat if it is at the expense of shocking a fellow rat.[17]

One of the fundamental demonstrations of our natural instinct to Bond with each other is a *will to give*. Rather than domination, our most basic urge is to reach out to another human being, even at a cost to ourselves. Giving to others—the urge to empathize, to be compassionate, and to help others altruistically—is not the exception to the rule, but our natural state of being. Our impulse to connect with each other has developed an automatic desire to do for others, even at personal cost. Altruism comes naturally to us. It is selfishness that is culturally conditioned and a sign of pathology.

In 2006 Felix Warneken of the Max Planck Institute for Evolutionary Anthropology and his colleagues decided to test whether a need to give is hardwired into our basic biological makeup and that of our closest relatives by designing a study with infant chimps and eighteen-month-old babies, who had not yet learned social behavior or group dynamics. The study took place at an ape sanctuary in Uganda where the chimps spent their days outside in the lush grass and evenings in their enclosure. Warneken asked a researcher unknown to both chimps and babies to place a wooden stick or a pen beyond his own reach in the enclosure, but easily within reach of the chimps or toddlers. Sometimes the stranger would simply stretch his arm through the metal bars, repeatedly attempting to get hold of the stick without success; at other times he would just look at it. In some instances the stranger offered a reward of a banana (for the chimps) or a toy (for the toddlers) when they helped him get the object in question.

Warneken figured that if the toddlers and chimps were responsive to the stranger's goal (getting the stick) they'd be more likely to hand the stick to him when he reached for it than when he just looked for it. On the other hand, if they were mostly interested in their own benefit in the situation, they would most likely help him get the stick when they knew they were going to get a reward.

The selfish-gene theory dictates that a toddler or chimp would help only if there were something in it for him, but experimental results confound that theory. Twelve of the eighteen chimps and sixteen of the eighteen toddlers reached for the stick and handed it to the stranger when he reached for it—regardless of whether they were given a reward. In fact rewarding their helping didn't seem to have any bearing on raising the rate of helping: both chimp and human babies automatically sought to help.[18]

Warneken and his colleagues then raised the stakes by placing the stick in a more out-of-the-way spot. In order to retrieve it, the participants had to engage in Herculean acts: the chimps had to run along an elevated raceway and the toddlers had to scramble over a blockade barrier. Even when the task caused discomfort or extreme effort, both groups demonstrated a selfless desire to help. Of course the two sets of babies might have been simply trying to please a more dominant individual (an adult human), so the German scientists decided to test whether chimps would put themselves out for other chimps.

First the scientists removed the possibility of personal reward by placing food inside a room with the door chained shut. The test chimps could see the food through their bars, and they could also see a stranger chimp trying to get the door open without success. The only way the stranger chimp could get to the food was if the chain were removed, but only the test chimp would be able to do this.

An extraordinary 89 percent of the time, the test chimps removed the peg on the chain to help out the strangers, even though they themselves would not benefit in any way.[19] These results are all the more remarkable because chimpanzees are known for being aggressive and dominating in general.

The automatic impulse to help appears to have its genesis in our internal programming to protect our young, as Joshua Green and Jonathan Cohen, two Princeton University psychologists, discovered when studying the brain signaling of people thinking about harm being done to others. The network of neurons in the brain that lit up during the process of witnessing another person being harmed was related to caring, the same neurons that fired when mothers were shown photographs of their babies.[20] The identical brain circuitry each of us possesses to help to care for our children is called upon to respond to the suffering of others. Caring about others, even strangers, is automatic and basic to our biology.

In fact a desire to help is so necessary to us that we experience it as one of our chief pleasurable activities. A team of neuroscientists from the National Institutes of Health and the LABS-D'Or Hospital Network in Rio de Janeiro discovered that receiving a large monetary reward and making a large charitable donation both activate the same portion of the brain, the mesolimbic reward pathway, a primitive system that gets aroused during eating or having sex. Getting something and giving something away are both highly pleasurable, but another section of the brain, the subgenual cortex or septal part, associated with bonding and social attachment, also fires when we make a charitable donation.[21] This would suggest that an impulse to do something altruistic is inherent in our need for connection.

As the evidence now shows, altruism is not socially induced, but hardwired within us, designed to be as essential and pleasurable as eating and having sex. We were made to feel good when we are doing good, and doing good appears to stem from our instinct to Bond.

James Rilling, a professor of anthropology, and Gregory Berns, a professor of behavioral sciences, both of Emory University, observed the real-time behavior of the brain during an altruistic act by employing functional magnetic resonance imaging (fMRI) scanners to record the brain activity of a group of women as they took part in a game called the Prisoner's Dilemma.

The Prisoner's Dilemma is a classic psychological game used to assess levels of cooperation between two people. In the most typical version the two participants are told to pretend that they have been arrested for robbing a bank, and they are placed in separate cells isolated from each other. At this point the police don't have enough evidence to convict either one, so the prosecutor pays each suspect a visit and offers him a deal. Each of them can rat on the other (called "defect" in the game) or remain silent ("cooperate" with the other). If only one of them betrays the other, while the other holds his tongue, the stool pigeon will go free while the silent one will get a maximum sentence of ten years. If both betray the other and confess, they will both be convicted and have to serve five years. If both remain silent, they will only be charged with possession of firearms and both will be released after only six months in jail.

The dilemma is that although each party is better off confessing than being the object of betrayal, that outcome is worse than if both remain silent and cooperate with each other.

Under these circumstances the best possible response is for both players always to defect, even though they would get a longer sentence than they would if it turns out that they both have cooperated, since by defecting the individual is better off no matter what the other prisoner does. It is the best possible overall response to any move by the partner.

In the iterated version of the Prisoner's Dilemma, the game is played over consecutive rounds with the same players. This really puts cooperation and altruism to the test, as the two prisoners are more likely to benefit if they work together than if they pursue their own self-interests.

In the Emory University version of the game, when the two players were asked independently to choose to cooperate with each other or to defect, each prisoner would receive a sum of money depending on both prisoners' choices in the round. Once again, the safest option, which will secure the biggest reward, no matter what your partner does, is to take the selfish option and defect.

Rilling and Berns were fascinated to discover that mutual cooperation—both players choosing to remain silent—was the most common outcome, even though defection was the safest choice.

What's more, when the partners cooperated both demonstrated activation in the caudate nucleus and anterior cingulate cortex, the same areas of the brain activated when people receive rewards or undergo a pleasurable experience.[22] Working cooperatively with someone else is literally its own reward. As a control, Rilling and Berns also examined brain activity when their participants were playing with a computer as the partner. In those instances the pleasure zone areas of the brain did not light up.

"Our study shows, for the first time, that social cooperation is intrinsically rewarding to the human brain, even in the face of pressures to the contrary," said Berns. "It suggests that the altruistic drive to cooperate is biologically embedded."[23]

Rilling believes that our built-in reward system reinforces our positive choices in helping—the more we do it, the better it feels—which in turn spur us on to help others even more. This response in the brain—feeling good about giving—represents the "genesis of a social bond," says Clint Kilts, co-investigator and associate professor of psychiatry at Emory.[24] An automatic *will to give* begins our process of seeking connection. Being selfless is ultimately the most self-serving option because it feels so good to give.

Other parts of our body are hardwired to make us feel good during acts of kindness and compassion, as Dacher Keltner, a professor of psychology at the University of California at Berkeley, has discovered. Our heart rate slows; our autonomic nervous system relaxes; we produce more oxytocin, the "love hormone" released by mothers after giving birth and breastfeeding. In Keltner's own experiments he has discovered that we understand the language of compassion through our most basic means of communication, touch and particular facial expressions, again suggesting that compassion and altruism have an evolutionary basis.[25] As a good deal of evidence shows, helping not only feels good; it also promotes health and even longevity.[26] It may even be an essential component of the contented life.

In the late 1930s Arlie Bock, the director of health services at Harvard University, conceived of the idea of taking the best and brightest from Harvard and studying them over time to determine which qualities in a person are most likely to make for lasting happiness. Besides satisfying his own curiosity Bock had big ambitions for his data, from which he promised to fashion a blueprint for "easing disharmony in the world."[27]

Backed by the department store magnate W. T. Grant, Bock and his colleagues from an impressive array of disciplines—medicine, anthropology, psychology, psychiatry, physiology, and social work—selected 268 young men at Harvard whom they had determined were the most promising, successful, and well-adjusted. The plan was to track their progress over many years in order to determine how exactly the lives of this bright bunch played out. For seventy years the men, told they were part of a chosen elite, were poked and prodded in every conceivable way, every body part measured and compared, from length of "lip seam" to scrotum size. Biological changes during physical activity were painstakingly chronicled. Psychiatrists submitted the young men to a battery of Rorschach and other psychological tests popular at the time. Social workers interviewed the men's relatives at length, uncovering even the most intimate of behavioral details, such as when they had stopped wetting their beds.

In 1967 the psychiatrist George Vaillant became the study's shepherd, monitoring the course of what the founders expected would be 268 success stories. In fact, however, many cases read like Shakespearean tragedy. Although a number of the group achieved extraordinary outward success—participants included the late president John F. Kennedy, a presidential cabinet member, a newspaper editor, a bestselling author, and four who ran for U.S. Senate—by age fifty a third of the men had suffered clinical mental illness. A good percentage had become alcoholics.

Many of those considered most gifted turned out to have disastrous or even pointless lives. One young man, the son of a wealthy doctor and an artistic mother, was singled out as exceptionally blessed. "Perhaps more than any other boy who has been in the Grant Study," wrote one researcher at the start of the study, "the following participant exemplifies the qualities of a superior personality: stability, intelligence, good judgment, health, high purpose, and ideals."

At the age of thirty-one the young man grew hostile toward his parents and eventually the world. Although the study lost track of him for a time, Vaillant and his colleagues finally located him, only to discover that he had lived nomadically, dated a psychotic girlfriend, smoked a good deal of dope, and dined out on a stock batch of humorous stories about his past before dying young.

Another young man, considered one of the most "bubbling and effervescent" of the group, took a succession of odd jobs and married and divorced several women before finally coming out of the closet and becoming a leader in the gay rights movement. Despite this newfound honesty with himself, he became a heavy drinker and at age sixty-four was killed after falling down his apartment building's stairs during a binge.

Bock was shocked by how his best and brightest were doing. "They were normal when I picked them," he remarked when Vaillant caught up with him in the 1960s. "It must have been the psychiatrists who screwed them up."

All the usual confident predictions about people's lives are defied by this sampling, as well as by another study managed by Vaillant, called the Glueck cohort. This study contains the flip side of the Harvard cohort: a group of boys from inner-city Boston, the offspring of poor and largely foreign-born parents, who were also followed for seventy years.

Although Vaillant is very careful not to overgeneralize what he has observed, he does note the same themes in both groups, rich and poor. Money and even a good start didn't guarantee happiness or success. Good luck didn't guarantee happiness. A particular personality type didn't guarantee happiness. What appears to be the most important factor is not how much difficulty you face in your life but your response to that difficulty.

As a psychiatrist Vaillant is particularly interested in "adaptations," or defense mechanisms: how a person unconsciously responds to stress, whether the stress is from physical pain, conflict of any sort, or even the unknown. As time wore on, the most successful among his cohort developed mature adaptations, such as humor or working out conflict constructively. However, among those living longest, one of the chief adaptive qualities that made for a long and happy life was altruism.

One young social misfit given to depression found his calling as a psychiatrist in midlife, inspired by the kindness of a health care worker during one of his bouts in the hospital. One tiny act of selflessness cleared his path, and he went on to have a highly successful life helping others.

On Tuesday, August 2, 2005, an Air France Airbus, attempting to land at Pearson International Airport in Toronto during a ferocious downpour, overshot the runway and crash-landed. Notified that most of the passengers had died, the Canadian governor general began issuing condolences to relatives of the 309 people on board. Once the smoke and rain subsided, it became clear that although forty people had sustained injuries, every single passenger had in fact survived. The plane had crashed near Highway 401, Ontario's main motorway. Hundreds of passing motorists had pulled over, rushed to the plane, entered the wreckage, and begun pulling out the survivors. Although two of the eight emergency exits were unsafe and the emergency slides did not work, the strangers smoothly coordinated efforts to get everyone out safely within a few minutes, before the plane burst into flames. Many of the evacuated passengers were picked up by drivers on the highway and driven to the Air France terminal.

Cases like the Air France crash upset the prevailing view of the mentality of bystanders that has existed since 1964, when, in a brutal attack lasting over an hour, twenty-eight-year-old Kitty Genovese was repeatedly stabbed while thirty-eight of her neighbors in Queens, either witness to the attack or within earshot of her screams, reportedly did nothing to help.[28] Genovese's murder and her neighbors' indifference provoked endless headlines and spawned a great deal of social psychological research into what has been called "the bystander effect": why people stand by and do nothing when another person is in trouble.

At the time the research concluded that if people are clustered together in any sort of group, there is a "diffusion of responsibility" they are less likely to help than they would if they were alone because each is waiting for someone else to volunteer to help first. Afterward witnesses are even prone to deriding the character of the victim to assuage their guilt at not lending a hand.[29]

The events that occurred immediately after the Air France crash challenge those assumptions. Hundreds of strangers interrupted their busy routine and risked their lives rushing into a crashed airplane to help a group of strangers they would never see again. Many even offered to drive them to the airport when any one of them might have been a terrorist who had deliberately caused the crash.[30]

The innate desire of people to help strangers was put to the test by a group of postgraduate students at Columbia University. For more than two months four teams of students had ridden the crowded Eighth Street A train from the 59th Street Station in New York City, bound for Harlem and the Bronx between 11 a.m. and 3 p.m. They deliberately chose that route because there were no stops between 59th Street and 125th Street, which meant that for seven and a half minutes a portion of the approximately 4,450 men and women who traveled the train during those hours would be a captive audience to an emergency situation.

Each team contained two student actors, a "victim" and a "model," who would come to the victim's aid if none of the bystanders did, plus two students to record the data as unobtrusively as possible. The victim came in two guises: either as a disabled person carrying a cane or a drunk reeking of alcohol and carrying a bottle in a paper bag. A minute after the ride began the victim staggered forward and collapsed on the floor. The model was to wait a predetermined amount of time before jumping in to help if none of the passengers did first.

In most cases the model didn't need to come forward: help was immediate and forthcoming for both types of victims. The victim with the cane received help 95 percent of the time, and even though some people were put off by the drunk, he nevertheless received help 50 percent of the time. The race of the victim didn't seem to matter; a black victim received help just as often as the white victims. The number of other people in the subway car also seemed to have no bearing on a bystander's willingness to come forward; in nearly two-thirds of the trials two, three, or even more Good Samaritans rushed to help.

The conclusion was inescapable and overturned most assumptions about the Genovese syndrome (which has since been discredited as a case of bad reporting). "People do, in fact, help with rather high frequency," the Columbia University authors concluded.[31] Even in that symbol of hard-boiled urban indifference, a packed New York subway, people helped each other out, no matter their race, the vast majority of the time. The subway Samaritan study and many real-life cases like the Air France incident suggest that when observing someone in trouble, most people will instinctively come to the person's aid. In many of us the desire to Bond is so strong that we enter a burning plane without thinking.

Samuel Oliner continues to study these kinds of people and their opposite numbers to tease out what distinguishes the "upstanders," the altruists, from the "bystanders," and why. He has interviewed 1,074 heroes of every variety—Christians who hid Jews from the Nazis, firemen who ran up the stairs of the Twin Towers on 9/11, Carnegie heroes and moral exemplars—in an attempt to pinpoint the qualities most essential to the altruistic personality. Although humans are innately generous and selfless, he has discovered, whether or not we display these qualities has a good deal to do with how we have been brought up to view the world. Our hardwiring gets reinforced when we see ourselves as one with humanity and view the world as a series of meaningful interactions.[32] Altruism also appears to be the natural extension of a loving, well-adjusted upbringing within a tight-knit community. Those who have been brought up with a sense of the Bond display it in their attitudes to their fellow man.

Oliner's studies offer several key findings. Our culture and upbringing either deprive us of our birthright or cultivate it. The ability to put ourselves in another's shoes is most developed in children who grow up with what Oliner and others refer to as the "altruistic perspective": *a strong sense of the social contract*. They are encouraged to become members of a community and follow its strict social code, to fulfill their shared obligations, sustain a good reputation, uphold friendship, and avoid social disapproval. They learn how to take their place in a community, if only in their families. As they grow they then learn to assist their society and not violate its rules or expectations.

The altruists also are able to view society as *all-inclusive* and to move beyond even large differences such as ethnicity and religion. Whether a non-Jew helping the Jews in Europe during World War II or a fireman rushing up one of the towers during the 9/11 attacks, the rescuers in Oliner's study often talked of *ethical and moral obligations to all humanity*: they viewed all of humanity as part of an indivisible Bond. They'd also learned that helping should occur without expectation of reward. Rescuers felt particularly empathetic toward others' pain and tended to view themselves and others in terms of their *mutual similarities,* regardless of religious, ethnic, economic, or sexual differences.[33] Altruists *cultivate friends among diverse groups*. They are able to see the big picture, beyond similarities or differences between people, to the common Bond.

Svetlana Broz discovered similar qualities during the war in Yugoslavia in the 1990s, after collecting ninety first-hand accounts from survivors, recounting ways that strangers had been "upstanders" with the courage to stand up to crimes committed against the innocent. They were, she wrote, "genuine examples of the goodness, compassion, humanity, and civil courage that continued to exist in these times of evil."[34]

Nancy Briton and Jennifer Leaning, researchers at Harvard University, conducted thousands of hours of interviews with people in twelve of the most conflict-ridden places on earth, including Bosnia, Afghanistan, and Cambodia. They discovered that upstanders had four primary virtues setting them apart, but most important was *a sense of common ground with people who were different*. They managed to find, in Briton's words, a "thread of similarity" with others, no matter how slight. These upstanders also took the Golden Rule to heart: *Do unto others as you would have them do unto you*. They treated their enemies in the way they wished to be treated.[35]

As Oliner and the others have found, the majority of altruists come from the healthiest, most cohesive background, in which the idea of holism and connection with all of humanity plays an important part. Bystanders and those who deliberately hurt others have had an upbringing that reinforced, on some sort of scale, the opposite impulse of social atomization: selfishness, prejudice, racism, and dishonesty. In most instances they did not have a close family life; as children many were ignored, deprived of love, or abused in some way. Selfishness was the natural result of a pathological, or at least defective, upbringing through excessive individuation.

Although a good upbringing may reinforce our altruistic hardwiring, why do we feel good about doing good? Is our motivation genuinely selfless, or is there some selfishness at the heart of it all? Most biologists argue that doing for others is usually carried out ultimately for selfish ends; what has been termed *social-exchange theory* claims that people behave unselfishly toward each other only if the benefits of doing so outweigh the risks. As the English philosopher Thomas Hobbes once remarked, anyone appearing to act unselfishly does so simply "to deliver his mind from the pain of compassion." According to this mind-set, we do nice things, basically, out of guilt or a fear of reprisal from our friends. The evolutionary biologist Michael Ghiselin put it more unsentimentally: "Scratch an altruist and watch a hypocrite bleed."[36]

Daniel Batson, a former psychologist at the University of Kansas who holds twin doctorates in theology and psychology, has run a number of experiments whose purpose is to look into the human heart to find out exactly why we do selfless things for other people. Am I helping purely to help my fellow man, or am I helping to feel good about myself? Or is it that I want to look good in other people's eyes? In his work Batson has managed to tightly control the experimental conditions so that he can distinguish true other-oriented helping from a host of alternative reasons, such as helping in order to win approval, enhance self-image, or avoid self-castigation. In some of his studies he made it either easy or difficult for his participants to avoid helping someone in order to find out what really motivated them.

Batson has managed to defend his empathy-altruism hypothesis against many challenges. In some twenty-five studies he has shown that altruism isn't sparked by social concerns, such as guilt, sadness, or shame.[37] Rather it is evidence of what he calls "the empathy-altruism hypothesis": people will help if they can put themselves in someone else's shoes.

In one study Batson and several colleagues made use of fMRI to study the activity in the brains of people who were observing video clips of patients undergoing painful medical procedures. When the observers projected themselves into the situation and imagined undergoing the procedure themselves, they increased their own distress, and those areas of the brain that are associated with pain became activated. But when they were able to focus their feelings on the inner state of the patient and his likely feelings, their own distress levels fell and their empathetic concern for the patient rose.[38]

Batson had similar results when he asked his study participants to take one of three perspectives when listening to a narrative about the difficult personal circumstances of "Katie," a young college student who had recently lost both parents. Those imagining themselves into the situation increased their own personal stress and blocked their ability to help. Those who were most effective at empathizing were able to step outside their own emotion and focus instead on Katie's likely feelings.

Batson's work suggests something important about the nature of the altruistic impulse. We are not prompted to help others when we imagine ourselves in their situation. Rather, we act on our natural compassion when we can tune in to another person's feelings, *move beyond the sense of self, and take the other's perspective.*[39] When people genuinely feel for another person's pain, rather than simply imagine their own pain in the same circumstance, they are prompted to act altruistically. Batson's work suggests that altruism requires an ability to move completely out of your own state of mind and into someone else's.

Oliner has found the answer to his life's question. When people are able to step outside their own sense of difference from each other they are capable of the most extraordinary good. The impulse is there from birth, but is realized more in some people than in others. Batson refers to this impulse as empathy, but Robert Cialdini, a former psychologist at the University of Arizona and the author of *Influence: The Psychology of Persuasion,* offers a more all-embracing explanation: *we help when we have lost our sense of individuality and step temporarily into a space of oneness.*[40] Altruism is the natural extension of what happens when you move out of your small sense of yourself and your own individuality and into the space between.

CHAPTER 8

TAKING TURNS

In the movie *A Beautiful Mind* the physicist John Nash, played by Russell Crowe, is sitting with some fellow graduate students in a bar in Princeton, circa 1948, with swing music blaring from the jukebox in the background.

"I'm not buying you gentlemen beer," says a preoccupied Nash, scribbling on a disheveled stack of papers.

"Oh, we're not here for the beer," replies one of his friends, as a striking blonde walks in with her brunette friends. All five of the young men are attracted to her, but immediately thrown into a quandary: Which of them is going to get lucky and end up with the blonde? One of Nash's friends quotes Adam Smith: "In competition, individual ambition serves the common good." From that perspective, one of the men notes, the best strategy is essentially *Every man for himself.*

"Adam Smith needs revision," says Nash, finally looking up. "If we all go for the blonde and block each other, not a single one of us is going to get her. So then we go for her friends, but they will all give us the cold shoulder because no one likes to be second choice. But what if none of us goes for the blonde? We won't get in each other's way and we won't insult the other girls. It's the only way to win. It's the only way we all get laid.

"Adam Smith said, the best result comes from everyone in the group doing what's best for himself. Incomplete. Incomplete!" Nash declares. "The best result is for everyone in the group doing what's best for himself . . . and the group."

With that, Nash bolts up and dashes out of the bar, nearly knocking over the blonde, stopping only briefly to thank her for being the muse to his idea, so he can get back to his room and scribble down the theory that made his name and eventually secured him the Nobel Prize.

The scene is meant to describe one of the most important moments in twentieth-century economics. Although given a good deal of poetic license, *A Beautiful Mind*'s simplistic version of the genesis of what became known as the Nash Equilibrium illustrates a basic principle of game theory, named by mathematicians since Nash. Game theory is a branch of math used to model and predict the behavior and strategies of individuals and groups of individuals in specific pressurized circumstances. As the name suggests, game theory is worked out with elaborate games that require certain decisions by everyone involved. Put someone in a tight spot and see what comes naturally to him. It is used to model strategic interactions—what an individual does in response to the actions of others when he is given a strictly limited set of choices and preferred outcomes. Most games also stack the deck by making it easier to go for the selfish option. By placing individuals in certain social dilemmas, game theory essentially measures the human capacity for grace.

The Nash Equilibrium is designed to work in a *noncooperative game,* in which each player makes his decision independently of the others. The point of the Nash Equilibrium, however, is that the choices you make should depend on what everyone else does. You choose the *best possible* position for yourself based on what everyone else is doing, and thus you all reach an equilibrium in which no one can improve his position given the choices everyone else has made. The movie version of Nash's aha moment isn't exactly correct, because if all your friends had chosen brunettes, the best choice you could have made (assuming blondes are the better choice) is to go for the blonde. Furthermore, in the movie version, one person does lose out, namely the blonde herself.

To a student of evolution, life essentially is a game. Although game theory was first used to predict strategy during the Cold War and then to describe economic behavior, the British evolutionary biologist John Maynard Smith realized its usefulness in 1972 in predicting which competing strategies help animal populations to propagate and survive. Game theory is now used for modeling in all the social sciences and in evolutionary biology. Biologists turn to game theory to determine how animals react in certain complex social situations.

Games strip away the vast complexity of animal behavior to its skeletal strategies. Biologists have devised a plethora of games with exotic titles—Hawk-Dove, Battle of the Sexes, Bourgeois, Transvestite, Sneaky, Scrounger—and used them to demonstrate which tactics lead to evolutionarily stable positions.

As a psychology undergraduate at Britain's University of Leicester, Lindsay Browning had been fascinated by a game called Hawk-Dove, named after polar positions in war, which demonstrated the need for stable cooperation between aggressive individuals and pacifists in any animal population when competing over resources. In the game, the opposing strategies are either aggression (Hawk) or yielding (Dove): Hawks always fight their opponents to the death; Doves simply demur in any showdown. A population of all of one or the other in any natural setting would not survive; the Hawks would go on killing rampages and deplete the entire population, and Doves would be so passive that any Hawk finding his way in would have a field day and quickly take over. The Hawk-Dove game demonstrates that the most stable state, from an evolutionary standpoint, is a mixed population of Hawks and Doves. But the question remained in her mind: at what point had Hawks learned to cooperate with Doves?

Browning wondered how to resolve the larger question of how cooperation ever evolved in the entire animal kingdom. If the fittest always survived, why would any living thing want to cooperate? A cooperative animal would help another, sometimes at its own cost, but this helping hand was open to constant exploitation by selfish animals. And, she asked, isn't natural selection supposed to favor the selfish? It was a bit like the contradiction of her own life: how to square her own ardent Christianity with her beliefs about evolution.

Browning had the good fortune to work with Andrew Colman, a psychology professor and one of the U.K.'s leading experts in game theory. She had caught Colman's eye because of her knowledge of computer programming, taught to her by her programmer father— an unusual string to a psychology undergraduate's bow. Browning's undergraduate thesis had concluded that cooperation had evolved as the strongest evolutionary strategy, but she wanted to explore it further. After graduating from Leicester she had received her doctorate from Oxford, gotten married, and had her first baby. She'd then received a small grant to return to Leicester as a postgraduate student to finish the work she and Colman had begun.

Browning mulled over the idea of creating a computer program that would test the evolution of cooperation through game theory. She was inspired by the iconic work of Robert Axelrod, a political scientist at the University of Michigan who had conducted a worldwide computerized tournament to solve the biggest unanswered question in game theory: the optimum strategy for repeated games of the Prisoner's Dilemma. This particular game was the perfect vehicle for studying the nature of cooperation because it revolved around the question of working together versus working alone. Should the two criminals cooperate or sell each other out? And, if you are trying to win with the highest score, why would you cooperate, which would likely mean a lower score?

Axelrod invited game-theory strategists from all over the world to participate, and they sent in computerized programs that could play two-hundred-round games of Prisoner's Dilemma against opponents. The winner was Anatol Rapoport, a Russian-born professor of mathematical psychology at the University of Toronto, who had a special interest in the psychology of war, peace, and nuclear disarmament. Rapoport won decisively with an ingeniously simple strategy, written in only four lines of computer code, which he'd dubbed "Tit for Tat."

On the first move the player cooperates with his opponent. On every subsequent move, the player simply copies the moves of his opponent in the previous game—but only for one round. If the other side defects, the first player defects in the next round; if the other side cooperates, the first player continues to cooperate until his opponent changes tactic, at which point, so does he. Following this strategy isn't simply kind; cooperation always yields the highest score.

Axelrod studied and published these first results and then ran a second tournament, with many challenges to Tit for Tat. Nevertheless Rapoport's strategy turned out to be unbeatable.

Axelrod tried his tournament a third time, this time attempting to give it real-life validity by asking programmers to program their players so that success bred a greater chance of survival: the more successful strategies in one round would be more prevalent in the following round, like a bunch of children who are better protected and cared for, and so end up with more progeny than their less safe and well-nourished opponents.

The final result of the third tournament was astonishing, particularly to evolutionary scientists: *niceness—cooperation—always ended up outperforming selfishness.*

Political scientists and religious leaders, marveling at these deceptively simple results, proposed that Tit for Tat offered lessons for all humankind. Metta Spencer, the editor of *Peace Magazine,* described it as an "exceptionally effective sanction, quickly showing the other side the advantages of cooperating." Rapoport's strategy offered nothing less than "a workable principle to use in real-life interactions."[1]

The outcome of the tournament forever changed Axelrod's view of cooperation, on which he expounded in a series of books.[2] Rapoport's player had adopted an "eye-for-an-eye" philosophy, punishing or rewarding the other player only in response to the other player's own behavior. Axelrod encapsulated the lesson to be learned from this in a few simple homilies: Always start by being cooperative, but return defection for defection and cooperation for cooperation. Nevertheless be forgiving and fair, and don't go on punishing your partner more than once after he has demonstrated selfishness.

No matter how ingenious Tit for Tat is in its utter simplicity, Lindsay Browning was left with several large questions about these results. Rapoport's program, in a sense, had been coded to have certain ethics: always to be nice or fair. She wondered whether there was an impulse for cooperation that was even more fundamental. She and Colman decided to test this theory by creating a program with a vast number of players who would play each other over a set number of rounds, after which they would "reproduce." At that point the binary code of the program's highest scoring players would be split in half. The first half of one "parent" and the second half of the other parent would be attached together to make a "baby." The outcome for each player would net each player a score, and those with the highest scores would get to reproduce more offspring than the losers. The twenty children with the high-scoring parents would then play each other. In this way the programs were to act as a crude approximation of natural selection; the fittest—that is, the most successful—game players would survive.

Colman suggested that Browning arrange it so that the players play a game called Battle of the Sexes. In this game the players are like a married couple who wish to go out for the evening. The wife suggests that they go to the ballet, but her husband wants to go to a prizefight. Both would rather go to their preferred venue than to their partner's choice, but both would prefer to go out together than alone to separate events. There are four possible scenarios: they can both go to the prizefight; they can both go to the ballet; they can both go to their partner's preferred venue alone; or each of them can go to his or her preferred option alone. Browning programmed the game so that the preferred option, both going out together at either venue, would net the highest score for the partner whose venue was chosen. As is possible with a computer simulation, which can run very rapidly, Browning programmed the game to play for 10,000 generations late every evening, after her baby Josephine was asleep.

One night Browning was awake at 1 a.m. analyzing the results when she noticed something peculiar. *The players seem to be taking turns,* she wrote in a guarded email to Colman. The highest scorers had evolved to take it in turns, getting their own way but then yielding to their partner's preferred option. Sometimes one player would let another partner have his way for one, two, or three rounds, and then the partner would return the favor—and for the same number of rounds. As soon as any pair began turn-taking by chance, they soon developed a perfectly coordinated effort, eventually settling into a stable relay of trading off with each other.[3]

At first this seemed impossible. She hadn't programmed the players to take turns or laid down any ethics in their decisions. All she had done was to program them to take advantage of the best strategy. The turn-taking had evolved spontaneously.

Browning ran the program repeatedly throughout most of the night, and then tried out her program on a variety of other games. Every round and every game produced the same response.

The following day Colman conducted a thorough search of the literature on cooperation, which confirmed what he'd suspected: no one had ever demonstrated that cooperation might have evolved through taking turns. Turn-taking had evolved spontaneously on its own as a Nash Equilibrium: it was not only the best strategy for the group; it was also the best strategy for each individual. They'd set up the computer program to perfectly mimic survival of the fittest in nature, and yet the survivors were the ones that were the most cooperative. It must be an innate drive in nature, Browning thought—a basic pulse of life itself. Taking turns was not only the key to cooperation and the primary glue of society, but also the most evolutionarily stable strategy. Everyone was best off simply by taking turns with each other.

This fit with her own strong Christian leanings: *Do unto others as you would have them do unto you.* When people want to be successful, they don't have to snatch the brass ring for themselves every time. The most successful strategy, the best response you can make for yourself, is purely and simply to wait your turn.

By a quirk of computer programming Lindsay Browning stumbled on another basic drive of our social Bond: the soul of any successful society is turn-taking, or *reciprocity,* a sense of fair play. The moment individuals begin to cluster in a group larger than the nuclear family they appear to evolve a strong, built-in sense of fairness. But as Browning proved, cooperation can be maintained only to the extent that individuals are fair with each other. Our survival depends upon our ability to give each one of us a turn, and the extent to which any society begins to fray relates to the deterioration in fairness and basic reciprocity. Both Axelrod's tournament and Browning's experiment revealed something even more profound: the understanding, deep within us, that we are best off in every way by taking a far larger and more all-embracing view of what constitutes self-interest.

It is astonishing how often the name defines the person. The Swiss economist Ernst Fehr, from the University of Zurich and affiliated professor at the Massachusetts Institute of Technology, has made his life's work the study of the economics of fairness. Fehr is his own kind of odd contradiction: an ex-champion wrestler fascinated by what makes for cooperation between people. He has conducted well-known studies demonstrating that people the world over not only are inherently generous, but also abhor unfairness—what he terms "inequity aversion."

Fehr has exhaustively tested his theory with the classic game-theory experiment called the Ultimatum Game. In this game volunteers are randomly paired, although never allowed to meet. The pairs are then split off into "proposers" and "responders." The proposer is given a sum of money, say $10, and allowed to offer the responder any amount of money, from $1 to $10, that he sees fit; the responder's job is simply to accept or reject the offer. If he accepts it, he will receive the designated sum and the proposer will keep the rest. If the responder rejects the offer, however, both leave empty-handed. This is a one-time-only offer, and both parties know this; hence the name "ultimatum." There is no possibility of holding out for a better deal. Furthermore, as the game is only played once, the two understand that there will never be any reprisal.

If human beings were innately selfish, it would make perfect sense for the proposer always to keep the lion's share and make the most derisory offer, and for the responder always to accept it, as something, no matter how little, is better than nothing. There is no social pressure to be generous in the game as identities are kept secret and the two will never interact again.

In practice this scenario rarely occurs with any pair in any society. The game has been played around the world and across many cultures under scientific conditions, and the outcome is extraordinarily uniform.[4] "If we pick two strangers with an anonymous exchange, there is a high probability that altruistic behavior will emerge spontaneously," says Fehr. The most common offer is 50 percent, and the overall average ranges between 43 and 48 percent.[5] Even though it means they stand to lose out personally, most people would rather share equally with people they haven't met and never will meet again.

Even more interesting, people tend to punish those who exceed the boundaries of fairness. Those playing the Ultimatum Game generally refuse any offer below 20 percent, so if they're playing with $10, they reject any offers of $2 or below.[6] Social scientists call this impulse "altruistic punishment," our desire to punish unfairness, even if it comes at a cost to ourselves. This suggests that within our social Bond with each other we have a strongly honed sense of fair play, which Fehr and his colleagues refer to as "strong reciprocity": the urge to cooperate with others if treated fairly and to punish those who violate the social contract of cooperation. So strong is this impulse that we are willing to cut off our own nose, so to speak, by refusing our own reward in order to punish transgressors who have taken more than they are entitled to. We would rather go home empty-handed than allow someone to get away with more than his fair share.

Research on the Ultimatum Game has mostly been carried out in university settings, with players largely made up of students. To find out whether this sense of fairness was true across the board in all societies, Fehr teamed up with an American anthropologist, Joseph Henrich in the Department of Psychology at the University of British Columbia, and headed out to remote places around the globe. The two scientists and their team of anthropologists began to test the Ultimatum Game in fifteen small-scale societies: foraging groups like the Hadza of East Africa; "slash-and burn" horticulturists such as the Aché, who live in the rainforests of South America; nomadic herding groups; and two small-scale agricultural societies.[7] In many cases the scientists had to replace money with some other currency, such as tobacco, when organizing the game.

No matter how the society was constructed, almost no one exhibited the Adam Smith model of inherent self-interested behavior, although some cultures displayed greater fairness than others. Unlike the university students, whose most common offer was a straight 50/50 split, the most common offers of proposers in these traditional cultures ranged from 15 to 50 percent, depending on how reciprocity was displayed in that culture.

Virtually all groups that had developed a society of any sort played the Ultimatum Game with a sense of reciprocity. The exceptions to the rule were the cultures that had not evolved socially beyond a cluster of individual nuclear families. The most common offer among the Machiguenga, an indigenous people in southeastern Peru, for instance, was to offer 15 percent and to keep the bulk of the money for themselves, but when they made these low offers, the offers were almost always accepted, no matter how derisory. This group is not clustered into a proper society and engages in very little cooperation or sharing beyond the family unit. The Machiguenga have a far smaller definition of "we," and so understand reciprocity only within the family. They do not appear to be concerned about the tide of public opinion, so they aren't shamed into offering more, particularly in an anonymous setting. The Hadza also tended to make some of the lowest offers, but also had high rejection rates; in real life these small-scale foragers do share meat, but hunters often hide their catch from other families, and they engage in frequent conflict. If their lives were a game of Battle of the Sexes, the Hadza haven't evolved past the first selfish stage of the game.

Individual offers tended to mirror certain societal mores. Members of the Au and Gnau tribes of Papua New Guinea most commonly offered more than 50 percent, but this offer was often rejected. In those cultures gift-giving confers status on the giver, and by accepting a large gift the receiver is made to feel inferior.

The whale-hunting Lamalera in Indonesia proved to be fairest group of all, far fairer than we in the West; two-thirds of the players offered a 50/50 split and the rest more than 50 percent. This society manages to make their huge catch only with the cooperation of many families. Afterward they meticulously divide the whale into designated sections and share it out among the community. For the Lamalera, sharing is tantamount to survival. Societies like theirs who play fairest in the Ultimatum Game are those most dependent on cooperation in their own lives.

Scientists have maintained that the human drive for fairness is a moral issue, developed only in human society as a "counterdominance" strategy, to "encourage"—by which they mean "force"—cooperation. Nevertheless, it appears that within any society a sense of fairness spontaneously evolves as a basic Bond of society, just as it did with Lindsay Browning's computerized game players. In the tests of the Ultimatum Game played with indigenous cultures, the more a society operated interdependently, the fairer the individuals acted toward each other. A sense of fairness in a group appears to evolve automatically as an inherent part of working together outside our immediate families. The roots of this impulse appear to run very deep and are primal in many living things, even monkeys, as Frans de Waal and his student Sarah Brosnan discovered.

Brosnan, a researcher at the Yerkes National Primate Research Center and Living Links Center at Emory University, devised an ingenious study for capuchin monkeys. This breed is well known for its cooperative behavior and strong social bonds, and capuchin females were chosen for the study because they tend to be monitors of fair treatment.

In this variation of the Ultimatum Game Brosnan placed pairs next to each other and trained one of the monkeys to hand over a small granite rock in return for a piece of cucumber. This is quite a concession, as animals generally are reluctant to give away anything. As soon as the trade was made, the researchers would then give the partner monkey the same size slice of cucumber if he played no actual part, or a grape, which, to a capuchin, is the far superior reward if he carried out the same work as his partner.

Both the monkeys who had made the original trade and bystanders witnessing the unfair treatment went berserk. They refused to engage further in any way with the humans, would not eat the cucumbers or grapes they had been rewarded, and in some cases hurled the food at the researchers.[8]

This study, which was published in *Nature,* suggests that a sense of fairness is an intrinsic part of the social Bond of our close relatives. Fehr believes that the fact that even monkeys reject unequal pay demonstrates that a sense of fairness is "very deeply rooted behavior."[9]

Brosnan believes that her study offers basic clues about how fairness evolved among humans.[10] In fact, the latest science shows evidence of an "it's not fair" spot in the human brain. Psychologists from Rutgers University enrolled forty male students, divided them into pairs and handed each member of the pair $30 apiece. Each partner then drew a ball from a hat, marked either "rich" or "poor." The "rich" participant received an extra $50 bonus, while his "poor" partner received nothing. The researchers then asked each participant in turn how he would feel about a further transfer of money paid either to him or to his partner, while monitoring activity within the participant's striatum and prefrontal cortex, brain regions thought to be involved in how we assess rewards.

Functional MRI scans revealed that brain activity was greater in the "rich" participants when they thought about money being transferred to their partner, but greater in the "poor" partners when thinking about getting a bonus themselves. In every instance, the players were less interested in making money for themselves than in rectifying financial inequality; in fact, as an individual's own financial rewards soared, so this neural activity quieted down. Every player mainly sought to close the monetary gap between himself and his partner.[11]

Like monkeys, people are so resistant to inequity that they are willing to give up their own material pay-off if that will bring about a more equitable outcome for all. In our hearts we're okay with a smaller piece of cake so long as we know it's going to mean that everybody else gets a slice. Fehr himself discovered this natural impulse in human beings when he investigated how groups of people behave when securing employment contracts. In one of his studies he gathered together a group of college students and divided them into a smaller group of "employers" and a larger group of "employees." He arranged that the employers would make a contract with their employees to provide a certain amount of effort, for which they'd be paid a set amount. However, the employees would be paid their contracted amount no matter how much effort they put in, so the employee wouldn't suffer any penalty if he didn't abide by his contract. Furthermore each employee was contracted to play the game only once with a given employer, and all identities were concealed, so there would be no stigma attached if the employee reneged on the deal.

If either party were entirely self-interested, the employers would be expected to offer the minimum wage, and employees would respond with only minimal effort. In practice this almost never occurred. Both parties were usually generous, and the more generous the employer, the greater the employee's effort. In fact the employers mostly assumed that their employees would work hard and so were bounteous in their wage offers. Nevertheless, only 26 percent of the employees delivered the full effort they promised.

In the next round of the game the employers were allowed to respond to their employees' effort by paying more than the promised amount for greater effort and punishing lower effort by paying less. In this case employers also showed an exquisite sense of fairness. More than two-thirds of the time they rewarded employees who did more than they were contracted for, and nearly half gave rewards to their workers simply for fulfilling the contract. This time, when contracts were not fulfilled, two-thirds of the employers meted out "punishment."

On the employees' side, with a reward for effort now in place, most did *more* than their fair share. Underachievement on contracts fell from 83 percent to 26 percent, and overfulfillment of contracts rose tenfold. Most significantly, allowing the bosses to reward or punish their workers according to effort increased the ultimate payoff to both bosses and workers by an average of 40 percent. This study reinforces the idea that, from our urge to connect, we have developed a strong internalized sense of fairness and we respond in kind.[12]

One of Fehr's interesting sidelines is examining how various hormones affect how we play games. He asks his players to inhale extra doses of either oxytocin, the "love hormone," which appears central to parental care, pair bonding, and social attachment, or testosterone, the hormone associated with greed-is-good selfishness. Fehr then uses fMRI scans to find out which areas of the brain are firing during the time the players are engrossed in games involving trust. Fehr likes to mix it up a little, giving oxytocin to men and testosterone to women. Although women produce testosterone as well as estrogen and other female hormones, their levels of testosterone don't vary as their levels of estrogen and other hormones do.

The studies using oxytocin yielded a predictable result. The men given oxytocin became more trusting of others and more likely to take risks. In fact they behaved as if they knew they could implicitly trust their partners. Even when the oxytocin group was double-crossed by their partners, they still showed evidence of total trust.[13]

In the testosterone study, Fehr's results undermined our preconceptions about how testosterone makes us behave. He gave some of his women players extra testosterone and others a placebo; although no one knew who had actually received the real hormone, Fehr recorded their views about whether they thought they had received testosterone. Then he had both groups play the Ultimatum Game. Those women who received an extra shot of testosterone actually played more fairly with their partners and made more generous offers, while the women who thought they'd received testosterone behaved more selfishly and unfairly.[14] The women who'd received the placebo were acting antisocially out of their own "negative assumptions" about testosterone, not the hormone itself, says Fehr.

Fehr produced an interesting analysis of his results. In his view, testosterone increases our awareness of status, and a mark of status in a social situation is generosity. *We think we will be better regarded by our peers when we make the graceful gesture.* Both these studies provide yet more evidence that our bodies are programmed to maintain accord, generosity, and trust.

Turn-taking entirely rests on the assumption that this impulse also exists in the person who is the object of our generosity, and that he will automatically return the favor. We immediately respond, as Rapoport's programmed players did, to any instance where we are in some way betrayed. Most of us possess a built-in scorecard that abhors a freeloader, with a corresponding need to punish those who take more than their fair share. Our abhorrence of unfairness is most evident in the fact that we are willing to punish transgressors of the social contract, even if doing so comes at our own expense.

This has also been proven in game theory with a game called the Public Goods, another standard in experimental economics. This game is designed to test how people behave when asked to contribute to something that could benefit the entire community, but at a price to themselves. It's a bit like asking people in San Francisco to voluntarily pay a sum of their own choosing in taxation toward maintaining the parks in all of California.

In this scenario a number of participants are given tokens, which are redeemable at the end for money. They're allowed to decide secretly how many tokens to keep and how many to put into a common pot. The experimenters then award some percentage of the total in the pot, 40 percent, say, to everyone playing. If each of four players is playing with twenty tokens and all four put in all their tokens, the experimenters will award 40 percent of eighty, or thirty-two tokens to each. Although each player forfeits all his own tokens, each ultimately benefits financially in the long run by putting all his money in the common pot.

The irony of the game is that everyone makes the most money when giving away all his own tokens, since the experimenters reward the most from the highest amounts within the pot. However, the Nash Equilibrium version of the game—the best you can do *under any circumstance, in response to what everyone else does*—predicts that the best response is not to put anything into the pot, but to keep all your tokens for yourself, lest you become a victim of freeloaders. However, this has almost never happened among the large number of Public Goods experiments carried out by Fehr and other social psychologists; most people add something to the pot, and on average give up half their tokens to the public good.[15]

This game, like the Ultimatum Game, can be run either as a one-shot or a repeat over a series of up to ten rounds. However, a very different scenario emerges during repeat Public Goods games. In that case, as Fehr discovered, the urge to give is initially enormous—most commonly, people begin playing by giving up to 40 to 60 percent of their tokens—but this generous impulse quickly abates so that, by the final rounds, nearly three-quarters of all people contribute nothing, and the rest close to nothing.[16]

Although at first glance it would appear that people are simply following their own self-interest, that isn't the explanation offered by the players. When interviewed later those participants who had initially been generous grew increasingly furious at freeloaders, who were contributing either nothing or less than the others. The generous players had retaliated with the only weapon available to them: they stopped contributing to the public fund. In other versions of the game, when players are allowed to fine the freeloader, albeit at a cost to themselves, they are more than happy to do so, even though they would benefit more individually by continuing to contribute to the pot.

Fehr ran these repeat Public Goods games two ways—either with or without the possibility of punishing freeloaders—and discovered that when people are allowed to punish those who are simply along for a free ride, cooperation in the game is maintained. Furthermore the biggest contributors turned out to be the biggest punishers. The ability to punish works even better when group members stay together throughout games, rather than swapping groups after every round.

When there is no possibility of punishment, however, cooperation quickly deteriorates, and the game, in effect, falls apart. We are willing to spite ourselves to stop others from violating our expectation of giving to the common good. It's a bit like a taxpayer, annoyed at the load of people on the welfare rolls, refusing to pay his taxes. Not only do we need to punish transgressors—it actually feels *good*.

Tania Singer of University College, London, who studies which situations prompt people to behave altruistically or selfishly, has discovered that our need for turn-taking in relationships is so ingrained that some of us are unable to feel empathy when we feel hard done by someone. Singer tested this idea in an intriguing study by examining the neural activity of thirty-two volunteers after they'd participated in a version of the Prisoner's Dilemma, in which pairs of players took turns sending their partners points, which would be exchanged for money at the game's end. Fair players reciprocated large numbers, while unfair players handed back smaller numbers, so that they'd receive the larger individual payout. Unbeknown to the participants, Singer had engaged two actors to play opposite the study's volunteers. The actors had been instructed to offer either excessively high or excessively low numbers, so that one played like a self-serving Machiguenga and the other like an ultracooperative Lamalera.

After the game the participants were placed inside an fMRI scanner while they watched their gaming partners, the actors, each receive a painful shock from electrodes attached to their hands. Both men and women showed neural evidence of empathy toward the fair players, but when the men saw unfair players getting shocks they showed evidence not only of reduced empathy but also increased activation in the part of the brain connected with rewards: they were actually *enjoying* the experience of revenge and favoring physical punishment of those who had got the better of them.[17] This suggests that our ability to feel empathy for others may in fact depend on whether or not the object of our attention has satisfied our inherent sense of fairness.

Freeloaders can come in many guises. One of the best recent examples of a society's sense of strong reciprocity occurred during the British miners' strike in 1985. When the Thatcher government announced the closure of twenty pits, which would mean the loss of some 20,000 jobs, miners across the country, particularly those living in the areas of the proposed closures, began to strike. As the strike dragged on for months, many of the miners, whose union money ran out, lived in extreme poverty without wages for an entire year, subsisting on soup-kitchen handouts and living in homes without heat. In a terrible irony, three teenage children of coalminers died that winter while foraging among piles of discarded pieces of coal to find some cast-offs to heat their parents' homes.

Not all members of the National Union of Miners had stayed behind the picket line, however. Many had straggled back to work after a few months; the workers in Nottinghamshire, whose pits were robustly successful, had voted not to strike and eventually created their own breakaway union, the Union of Democratic Mineworkers.

The scabs, as strikebreakers are known, were considered the ultimate freeloaders, who undermined the success of the strike. By agreeing to work they took more than their fair share, and for years the impulse burned in many miners to punish their defection. Although more than two decades have passed, Neil Greatrex, the president of the UDM, is just one of many who continues to feel animosity from those who had remained on strike. His father refused to talk to him for six years; his wife and children were constantly threatened. His next-door neighbor, a policeman, was obliged to put a sign in his window, "Greatrex lives next door," after growing tired of fixing windows broken by flying bricks. Hatred of the scabs reached its apotheosis in the autumn of 1984, when a taxi driver named David Wilkie, who had been driving a nonstriking miner to work, was crushed to death in his car by a concrete post deliberately tossed by two of the miners in the picket line from a bridge above the road.

The power of strong reciprocity to hold together a society or lead to its collapse was evident in the BBC's prison study, *The Experiment*. Eventually the prisoners staged a breakout, and the authority of the guards collapsed. After the coup everyone, guards and prisoners, spontaneously agreed to form a system of greater equality, what they termed a "self-governing, self-disciplining commune."

The idyll didn't last. When the group began doubting their ability to enforce the commune, and numerous individuals did not create sanctions for rule-breakers, the organization of the group began to fall apart. Some members began to plan a new coup in which they would take over, redefine the line between prisoners and guards, and establish more authoritarian order. They even requested black berets and sunglasses to reinforce their image of bad-ass authority. Those in the commune didn't fight back and were in such disarray that they too agreed to a system of tyranny again. At this point, fearing a repeat of the Stanford prison experiment results, Haslam and Reicher ended their experiment.

The ultimate problem with the BBC's utopian ideal was that it did not create a culture of interdependence defined by *strong reciprocity*. With just a few freeloaders and rule breakers the entire community broke down and could be held together only by force.

The consequence of acting against our deepest nature by pursuing a paradigm of selfishness and ignoring our basic drive for connection has been grave. Two British epidemiologists, Richard Wilkinson and Kate Pickett, spent more than thirty years painstakingly examining why certain human societies live longer and healthier than others, the results of which are contained in their remarkable book, *The Spirit Level: Why More Equal Societies Almost Always Do Better*.[18] After researching the social conditions of virtually every Western country, Wilkinson and Pickett discovered an astonishing statistic that ran through every country they'd examined: the more unfair any society—which is to say, the more economically unfair and hierarchical—the worse off *everyone* is—both rich and poor—in terms of virtually every social problem. In countries with giant income disparity between the very rich and the very poor, *both the most affluent and the very poorest* suffer from higher rates of ill health, crime, mental illness, environmental problems, and violence.[19]

Western countries in the main are at their most unequal in history, where our sense of taking and giving has been replaced with taking whatever you can get for you and yours alone. The United Kingdom, the United States, and many countries in Europe, with their vast differences between rich and poor, are among the worst off in virtually every social indicator compared with countries like Japan and Sweden, which have less wealth disparity in the population. Although one of the wealthiest countries in the world with half the world's billionaires, America has far and away the highest level of all social problems—crime, lack of education, mental illness, suicide, disease of all varieties—of twenty countries, and the United Kingdom has the third highest. Although one in every thirty-nine Americans is a millionaire, one in seven, or 39.1 million Americans, lives below the poverty line.[20]

One-quarter of all people in America have been diagnosed with mental illness—the highest figure among the most advanced developed countries—compared to less than one in ten in Germany, Japan, and Spain. Although the United States spends nearly half the entire world's expenditure on health, and has only 5 percent of the world's population, the fact remains that a baby born in the United States has a 40 percent higher risk of dying during the first year than a baby born in Greece, one of the poorest countries in Europe, where the population makes half as much on average and the country spends half as much on health care. Furthermore that Greek baby can expect to live 1.2 years longer than the American. Fairness, like belonging, appears to be necessary to our survival.

Fairness indicators have nothing to do with the amount of governmental expenditure that aims to redistribute wealth and equalize society. For instance, the state in the United States that reports the lowest level of social problems is New Hampshire, but it also has one of the lowest public expenditures of any state. New Hampshire simply does not have a huge disparity between the income of its richest and poorest inhabitants.

Our reaction to inequity has nothing to do with a need for sameness—for a socialist-style, across-the-board equality. Throughout history the fact that there is a wealthy group of individuals at the top of a society has not automatically made for revolution. Poorer levels of society are usually prompted to rise up in rebellion only when conditions are manifestly unfair, such as when food is deliberately made scarce. In the wake of the worldwide financial crisis of 2008 the fury that most ordinary citizens felt toward bankers and traders had nothing to do with income resentment but a deep and compelling sense of unfairness that investment houses like Goldman Sachs still paid record bonuses after the recession they had helped to create caused so many others to lose their jobs. In Britain Sir Fred "the Shred" Goodwin, former chief executive of the Royal Bank of Scotland, unapologetically paid himself a £700,000 pension (about $1.05 million) despite the bank's sustaining, under his stewardship, the largest corporate loss in history, requiring a £24 billion government bailout. Shortly thereafter aggrieved citizens attacked his Edinburgh villa and smashed his Mercedes S600. A statement sent to the *Edinburgh Evening News* read, "We are angry that rich people like him are paying themselves a huge amount of money, and living in luxury, while ordinary people are made unemployed, destitute, and homeless."[21]

Our sense of unfairness emerges when the most fundamental needs we have—to belong together, to fuse in body and mind, to give generously of ourselves, to take our turn—are thwarted, when the promises we make to each other to wait our turn, to take our fair share, are broken. In most developed communities today it is every man, woman, and child for himself.

After publishing his groundbreaking *Bowling Alone,* which exposed the breakdown of community life in America, the political scientist Robert Putnam ran a comprehensive survey of 30,000 people to investigate the effect of ethnic diversity on trust and civic engagement in modern America. He was dismayed to find that the more ethnically diverse an area, the less likely the citizens were to engage with each other or indeed with members of their own ethnic community. What's more, he discovered similar statistics in other countries: the more ethnically diverse a community, the lower its societal trust.[22] In fact the more ethnically different people are, the more they're likely to cheat at Prisoner's Dilemma and Ultimatum games.

This suggests that trust requires a sense of holism, a larger definition of "we"; as soon as we designate someone as the "other" we reinforce our own sense of separation and no longer feel it necessary to play by the rules. Eric Uslaner, a political scientist at the University of Maryland, studied trust in societies around the world for his book *The Moral Foundations of Trust*.[23] He found that the more unequal a society, the less trusting its citizens. "Trust," he says, "cannot live in an unequal world." Competition, in a sense, kills the Bond.[24]

We live now with the uneasy feeling that something profound in our lives is missing, a longing we cannot quite define. We are experiencing a disengagement from our birthright, our need to live in holism. We're left with something far worse than a simple case of unfairness, the vague sense of an important connection that has been broken: the evidence, everywhere we look, of unrequited humanity.[25]

PART III

RECOVERING THE BOND

Music is the space between the notes.
Claude Debussy

CHAPTER 9

SEEING THE WHOLE

When the three eighty-foot waves of the December 2004 tsunami struck Bon Yai beach on South Surin Island in Thailand, the Moken tribe, a small nomadic community of fishermen, witnessed the destruction of their village and the instant deaths of 24,000 people from a safe haven on one of the island's highest hilltops. Tribal elders had alerted the entire Moken tribe of two hundred, and all save one, a handicapped boy, were successfully evacuated well before the waves struck. By the time the tsunami swept north, reaching the Andaman and Nicobar islands and southern India, all 250 members of the ancient Jarawa tribe living on the otherwise deserted island of Jirkatang had already fled into the Balughat forest. After ten days of surviving on coconuts they emerged unharmed.

All members of the other four indigenous tribes of the Indian archipelago of the Andaman and Nicobar islands—the Onges, the Great Andamanese, the Sentinelese, and the Shompens—were also said to have had a foreboding of the tsunami, even though ordinarily they would have been out fishing on the seas. When an Indian helicopter circled the island, scanning the region for survivors, a naked Sentinelese, affronted by the needless intrusion, grabbed his bow and lobbed an arrow at the chopper.

When asked how they knew the tsunami was coming, an elder of the Jarawa tribe shrugged. It was obvious. One of the small boys of the tribe had felt dizzy. The level of the creek near their village had suddenly dropped. One of his tribesmen had noticed a few small differences between the way one wave swelled compared to another. They'd noticed an unusual restlessness and scuffling among the smaller mammals, a slight alteration in the swimming patterns of the fish. As a child, the elder had been taught to pay attention to these subtle signals. They warned of tremors from the earth and the sea about to hurl forward with fury. The elder understood that these were signs, that the sea and the earth were "angry" and that his people should flee to higher land.[1]

One of the worst-hit areas of the tsunami included Yala National Park, Sri Lanka's largest wildlife preserve, where tidal waves flooded up to two miles inland. Yet according to Ravi Corea, president of the Sri Lanka Wildlife Conservation Society, of the hundreds of animals at the preserve, only two water buffalo died. Hundreds of elephants, leopards, tigers, crocodiles, and small mammals hid in their shelters or escaped to safety.

The remarkable survival of wild animals and indigenous people has been variously attributed to an exquisitely attuned sense of hearing, a "seismic" gift that allows them to sense the vibrations of the earthquake, or an ancient understanding of subtle changes in the wind and water. "They can smell the wind," says Ashish Roy, a lawyer and environmental activist, about the native islanders. "They can gauge the depth of the sea with the sound of their oars. They have a sixth sense which we don't possess."

But another possibility is something even more extraordinary: an enormous difference between the way they see the world compared to the way we see it. A year before the tsunami hit, Anna Gislén, an ophthalmic biologist at Lund University in Sweden, was intrigued when a colleague told her that the sea gypsies, as the Moken are referred to by outsiders, had the extraordinary ability to collect obscure food from the sea floor. They could even distinguish small brown clams from brown stones underwater without the use of visual aids. Such a feat wasn't usually possible, even with goggles, as human beings are poorly adapted to see underwater. When we are out in air, two-thirds of our eye's refractive power is due to the curvature of the corneal surface, an advantage that is lost when we are swimming underwater.

Gislén headed off to Thailand's Surin Islands and began conducting underwater tests on Moken children, comparing their visual ability with that of European children on holiday in the surrounding areas. What she discovered confounded most ideas about human biology. Ordinarily, when we're immersed in a blurry environment, such as water, our eyes don't attempt to focus, which is exactly what Gislén observed with the European children she studied. The Moken children's underwater acuity, however, was more than twice as sharp as the Europeans'.

A Moken child learns to swim before he learns to walk. He's taught to slow his heartbeat when underwater in order to stay there longer. The Moken children train their eyes to control the "accommodative response" to the blurry underwater environment by constricting their pupils to a diameter that is 0.7mm smaller than the Europeans can in order to improve depth perception—"the same process that improves focal depth if using a camera with a smaller aperture," Gislén noted.[2] This minor accommodation improves vision so much that they are able to locate small clams and sea cucumbers, even at a depth of ten to thirteen feet underwater.

In every regard the Moken have learned to see with a better pair of eyes. They have turned their eyes into cameras, changing the aperture at will. They are able to notice detail and connections that most of us can no longer see.

They see into the space between things.

We have lost our sense of the Bond, but our loss is not irrevocable. We can recover wholeness in our lives and recapture our sense of the connection between things, but doing so requires a very different set of rules from the ones we currently live by. To live the Bond is to surrender to nature's drive for wholeness and to recognize the whole in every aspect of our daily lives. We have to ask ourselves some fundamental questions: How should we view the world as something other than a place for just ourselves? How should we relate to each other, if not competitively? How might we organize ourselves in our neighborhoods—the immediate tribe around us and our smallest group outside the family—to be mutually supportive rather than competitive?

How do we move beyond *I win, you lose*?

We need to perceive the world differently, relate to others differently, organize ourselves—our friendships and neighborhoods, our towns and cities—differently. If we're not to be separate, but always attached and engaged, we need to change our fundamental purpose on Earth to something more than one based on struggle and domination. We must look at our lives from an entirely different perspective, a larger vantage point, so that we can finally see the interconnection. We must change the very way we see the world so that we see as a Moken sees, not to forecast tsunamis, but to notice the connections that tie us all together.

Most of us in the developed world live with an atomized view of the world and lack the ability to perceive the subtle connection between things. We've developed a peculiar form of tunnel vision that concentrates on looking for the individual thing. That thing is usually an object, or even ourselves, but it can also be the central protagonist or even the central point of the story. We train ourselves to perceive the world as a load of individual and separate objects and categories of objects or separate ideas. Attending to the separate object entails seeing the world entirely from that object's particular vantage point: categorizing the object, applying rules to it, and working out causal relationships only in terms of the object itself. We're looking for the central element in the drama, the central object as separate from its background, and on that we train all of our attention. We cannot see the wood—the forest—for the trees.

We have forgotten how to look. We miss the subtle connection, the peripheral idea, the slightest of changes in the wind that lead us to the inescapable conclusion that a tsunami is about to hit. Even the Moken who were out in their boats before the tsunami struck knew to go to deeper waters and to stay away from the shore, unlike the Burmese fishermen nearby, who perished. A Moken greets the news of their demise with a knowing nod: "They were collecting squid. They saw nothing. They don't know how to look."[3]

We have seen that our most fundamental need is always to seek connection and unity and to move beyond individuality, yet when we look at our world all we see are separate and unrelated, individual things. Our most basic impulses about ourselves run counter to how we presently see and interpret our world. By learning to see like a Moken—to see the space between things—we may learn to recognize the connections that were always there but remain invisible to the Western eye: the connections that tie us all together. We will begin to recognize what is most invisible of all: the impact of ourselves on others and on our surroundings. We will notice the ripple effect of every action on an entire chain of being—the living things, the natural world, the friends along the network, the members of our community, the people from other countries whom we benefit or harm by what we do. Just as a Moken sees a chain reaction of events from the stirrings of a bird or the swimming patterns of a fish, so we may be able to see into the space where difference fades away and we find our common ground.

The lesson of the Moken runs far deeper than hunting for clams or surviving a tsunami. It shows us that when human beings look at the world out there, not all of us see the same thing. Our individual cultures teach us how to look and what to see. By recognizing this, we can begin to take a larger and more all-embracing look.

Imagine two international students, one from Japan and the other from the United States, at the Louvre, fighting the crowds to stand in front of the bulletproof, climate-controlled cage housing the *Mona Lisa*. Both are instructed to describe the painting. The American immediately focuses on the woman herself and her most celebrated mysteries: her identity, now assumed to be Mona Lisa Gherardini, the third wife of wealthy silk merchant Francesco di Bartolommeo di Zanobi del Giocondo, and the source of her enigmatic smile. He is drawn to the painting's very epicenter: the triangle of features encompassing the woman's eyes and lips. He notices the *sfumado* registering the shadows on the outer points of her eyes, painted in deliberately to obscure the emotion of the sitter, and the odd smile playing on the left side of her mouth. Every other part of the painting seems stuck on like an afterthought and may as well disappear. No matter how long he stares at it, the American cannot integrate the wood—the rest of the painting—for the single tree—the figure in the foreground of the frame.

To the Japanese student, however, the painting represents a metaphysical statement about the cosmos: the connection between humanity and nature. His eyes flicker back and forth between the figure and the background landscape, noting details such as the finery of the stitching in the woman's black veil or her curves echoed in the extraordinarily elaborate background landscape, with its winding paths and rivers and glimpse of the bridge at Buriano. He pauses to ponder the woman's complete lack of jewelry, uncharacteristic for that period in history. For a good half hour he regards the painting through his fingers at different angles, using his paper museum guide as a kind of ruler, to work out the meaning of the violations of perspective and the variation in the size of Mona Lisa's hands.[4] Through his eyes the woman's features entirely disappear into their surroundings. There may as well not be a person in the foreground, so little does she stand out for him. Mona Lisa cannot be freed from her context. The Japanese student literally cannot see the tree for the wood.

In the difference between what these two students see lies a good portion of Richard Nisbett's career. Nisbett, a professor of social psychology at the University of Michigan and the author of *The Geography of Thought: How Asians and Westerners See Things Differently and Why*, has made a life study of cultural influences on methods of thinking.[5] Nisbett contends that thought processes and even perception itself are not universal, but a cultural phenomena. People do not perceive the world in the same way in different parts of the world; they don't even see the same thing. Nisbett's extensive work in this area, which he terms "the geography of thought," clearly shows that different cultures develop very different styles of thinking. It starts with the very way we're taught to see.

For most Americans older than fifty, *Dick and Jane* was the first book they read, and it described a perfect American fantasy, recording the comings and goings of the rosy-cheeked brother-and-sister duo, their little sister, Sally, and their black-and-white spaniel, Spot. They inhabit a *Leave It to Beaver* world, with a dad who sports a suit, even on weekends, and a mom who wears a pretty pastel day dress, even in her own kitchen. The narrative always focuses on one of them getting up to something: "Look Jane. See Dick. See Dick Run," as Dick tears across the lawn.

According to Nisbett, books like *Fun with Dick and Jane* (and their British counterparts, *Janet and John*) taught children not just how to read, but how to *see the world*. In Dick and Jane's world, and ultimately in the Western world, a child learns to zero in on the individual. What we do by ourselves—what we get up to, how we feel, what we achieve by ourselves—is central, the very point of our existence. Our parents and schooling emphasize individual excellence over all else. We learn that we are the subject and everything else is the object. We see the rest of the world entirely as it relates to us. We are taught—encouraged—to be separate, the central point of the story.

From the moment he is born a Western child is taught to work toward independence. Early training from the crib teaches us how to think, but also, in a sense, how to *be*. We learn that autonomy is most important.

As Nisbett notes, Western babies are encouraged to sleep alone and to move as quickly as possible toward independence in their thinking and choices. Mothers introduce the world in terms of objects to be labeled and choices to be made: Ham or eggs? The red pen or the blue pen? *Barney* or *Sesame Street*? The late anthropologist Edward T. Hall called this kind of thinking the outcome of a "low-context" society, by which he meant that our identity is independent of our context. We view ourselves as an unfettered free agent; take you or me out of our society, plant us somewhere else, and we will still be the same person. These overriding, atomistic ideas—that our identity is autonomous, that we are the master of our universe—inform how we interpret and relate to the riot of sensation and stimulus coming at us at every moment.

East Asian children learn to read with very different ideas. In one of their first reading books, a little boy sits on the shoulders of a bigger boy: "Big brother takes care of little brother. Big brother loves little brother. Little brother loves big brother."[6] An East Asian understands himself only in his relation to the whole, whether the whole is represented by his family, his neighborhood, his culture, the Tao, or even his sense of consciousness.[7] As children, East Asians (and also many indigenous cultures, like the Moken) are brought up with such a strong sense of connection to others that they can see the self (and objects) only in relationship to their context. Because Easterners define their world so differently, they learn to view it with a different pair of eyes. In the East a child learns about the relationship and its primacy—that he and someone else are a unit, an indivisible Bond.

Consequently, according to Nisbett, Eastern cultures actually think differently than we do in the West. Traditionally the Chinese (whose culture has influenced many others in the East) have learned to understand things only in regard to other things. They see life in relation to a field of forces and understand matter in the universe not as a set of discrete objects but as mutual, continuous, and interpenetrating. The world, to the Easterner and to native cultures, is in flux, ever mutable and in the process of becoming. The Eastern or indigenous mind has learned to see the world far more holistically from the moment it is conscious.

Indigenous people, such as Native Americans, also learn to take in the totality of the physical and emotional landscape. "Seeing involves mentally experiencing the relationship between tangible and nontangible things in the world and in the universe," says Donald Fixico, a Seminole and Muscogee Creek who has studied the differences between Indian and Western "linear thinking."[8] To a Westerner, "thinking Indian," as he describes it, is a bit like being in the middle of a hallucinogenic experience. It blends the visible and invisible, the present and the past, the dreamer and his surroundings. All of those relationships, past and present, color what a Native American sees.

The stories we tell ourselves about how the world works ultimately govern what we perceive. After a while we see only what we are taught to see. This is partly due to a mechanism in the brain referred to as "kindling," discovered and named by a neuroscientist, Graham Goddard, who first accidentally discovered the phenomenon in 1967 after a surprise outcome in an experiment with a set of laboratory rats. Goddard was fascinated by the neurobiology involved in learning and wondered whether electrostimulation might speed up the process. In his experiment he electrically stimulated the brains of a group of rats every day, enough to provoke a seizure, to see if this had any effect on their ability to learn. After some days he noticed something completely untoward: the rats began having seizures even when the electrical current and charge applied to their brain were far too low to provoke a seizure. Somehow he'd trained their brains to become epileptic.[9] As a result of Goddard's work, modern neuroscientists believe that, like a coal fire that blazes more easily when ignited with small pieces of wood, so pathways within the nervous system become sensitive to certain connections if they are reinforced early on, after which they begin to occur more easily or with greater frequency.

The theory of kindling has been applied to bipolar disorder and depressive illness; it's now believed that the more depressed someone has been in the past, the more easily he gets depressed in future. From our understanding of brain plasticity, we also realize that kindling is a feature of perception.[10] Over time the pessimist can see only the negative in any given situation, and the optimist can see only the positive. For us in the West, so used to picking out individual things in our world, the center of the picture is now all we can see. We are always looking for the star of the show.

In a fascinating series of studies Nisbett and his team in the Department of Psychology at the University of Michigan revealed stark differences in the way Westerners and Easterners perceive the world. Working with a colleague at Hokkaido University in Japan, Nisbett gathered together two groups of students from the two universities and showed them twenty-second videos of underwater scenes. After viewing the film twice, each participant was asked to report what he saw.

The Americans invariably began describing the scene with the objects in the center: the fish. To the Japanese the context was most important; they saw the field itself: the color of the water, the plants, the ocean floor.[11] They were even sensitive to the inner lives of the fish and more likely than the Americans to describe what they interpreted as the fish's emotions.[12]

When Nisbett altered the videos slightly and then showed it to both groups of students, the Americans were most likely to detect the changes if they related to the central object, and the Japanese were most likely to see the changes if they concerned the background environment.[13]

Nisbett has even found that Easterners and Westerners have developed different ways of using their eyes to take in their surroundings. When he monitored the eye movements of a group of American and Chinese participants while they viewed batches of photographs, each with a single object, such as a tiger, in the foreground against a complex background, the eyes of Americans quickly fixed on the tiger, whereas the eyes of the Chinese flitted from one point in the background to another. The Chinese made use of far more rapid intermittent eye movement than the Americans, but also required far more time to take in the entire image. From their upbringing, they'd learned to attend to the whole far more than did the Westerners. When viewing the same scene the two cultures actually saw something quite different.[14]

Nisbett then asked groups of Japanese and Americans to take a photo of one another. The Japanese would always photograph the entire scene, with the whole person framed to be relatively small against the entire background, whereas Americans photographed the person in close-up.

What all this adds up to is that our philosophy of the world and how we see ourselves in relation to it govern what we actually see. In the West we are so busy picking apart what we see, looking for the individual thing rather than the Bond, that oftentimes we miss the vital connection right in front of us.

*   *   *

The British illusionist Derren Brown has a favorite party trick. While holding a map, he approaches a stranger on the streets of London and asks directions to St. Paul's Cathedral. In the middle of his inquiry, actors dressed as workmen walk between him and the stranger, carrying a giant billboard that momentarily obscures them from each other. During this diversion Brown carries out a quick sleight of hand: he disappears and is replaced by another actor holding a map, who carries on asking for directions as though he were Brown. At first Brown makes the switch only with someone who resembles him—a tall, dark-haired, wiry male in his late thirties—but as time goes on, with subsequent passersby, Brown gets more and more emboldened, switching with a gray-haired man, then a bald man, then a black man, and finally even a woman. No matter how dissimilar the person with whom he swaps, at least half of the people he approaches for directions don't notice. In fact they don't even recognize Brown, a popular television personality, or indeed observe that the billboard interrupting their conversation sports a giant picture of him. Brown could be anyone. Although they're looking at him, he hasn't registered on their psychic landscape.[15] They aren't really paying attention. They suffer from what psychologists prefer to call "inattentional blindness."

In order to process the riot of sensation and information around us, we have to be selective in our attention. We have learned to focus so tightly that we see far less of the world than we think we do. When our attention becomes diverted or focused on an object or task, we often fail to see what is happening right before our very eyes. Although we may perceive everything in our field of vision, we will only process what we place our attention on, and we in the West have learned to strictly limit that field of focus. We may think of ourselves as a camera, absorbing everything in our sight, but even extraordinary events don't necessarily capture our attention if we are absorbed in something else.

Christopher Wickens, an aviation psychologist at the University of Illinois at Urbana-Champaign, has studied how pilots in flight simulators perform when "heads-up" displays of information about critical flight information, such as airspeed and altitude, are superimposed onto the windscreen. These pilots are often so engrossed in the flight information that they often miss the unexpected—including an airplane on the runway—even if it is directly in their field of vision, and land right on top of it. Novice pilots, who aren't familiar with landing, don't suffer from these bouts of inattention; they see the obstructions each time. Because it is all new to them, they attend to the entire scene with complete and rapt attention.[16]

Selective looking occurs so frequently among lifeguards that it has spawned a host of lifeguard guidebooks about how to combat a common problem: lifeguards anxiously scanning the pool and monitoring all the swimmers around it frequently fail to notice bodies on the bottom of the pool, particularly if the body is directly below the swimmers.[17]

Two psychologists, Arien Mack of the New School for Social Research in New York and Irvin Rock of the University of California at Berkeley, first coined the term "inattentional blindness" after observing their study participants become so engrossed in a display they were told to watch that they failed to notice a bright red rectangle, even when it appeared dead center in their field of vision.[18] Mack believes that although we have no conscious perception of events around us that we are not paying attention to, the brain nevertheless carries on noticing and registering this raw information, particularly if it has some meaning to us. Mack and Rock found that even when study participants did not appear to register certain information, they had managed to notice and store it, because they were able to make use of it later in the test. Although we may not be consciously aware of things that occur if we're not paying attention to them, they are nevertheless filtered through our perception. We see all the wood around us, even if we've been taught to pay attention only to the biggest tree.

For the longest time scientists believed that our unconscious screening process, termed "latent inhibition," of the constant stream of sensations and stimuli to which we are exposed every day is there to prevent sensory overload. Supposedly, reduced levels of latent inhibition led to psychosis: schizophrenics weren't crazy per se; they were just people who had been exposed to too much all at once.[19] However, the revised view is that a poor ability to screen stimuli from conscious awareness may actually be a sign of genius, indicative of a superior ability to engage in creative thinking. The Harvard psychologist Shelley Carson and her colleagues discovered that people with a lifetime of creative achievement have significantly lower scores in latent inhibition than other people. The most eminent creative thinkers were seven times more likely than others to have low ability to screen out stimuli.[20]

Carson even found that highly creative individuals have certain neurobiological similarities to schizophrenics, in the ways they think and their access to more unfiltered stimuli. What distinguishes the lunatic from the poet, according to Carson, is creative intelligence. An inability to screen out sensory overload makes you crazy unless you are highly intelligent, in which case you can put this information to good use. The difference is that the highly creative person doesn't get overloaded, but uses the information to put together ideas in a new and arresting way. By "creative," she means someone who possesses a natural openness to new experience and a conscious desire to avoid tuning out the seemingly irrelevant. A creative thinker, like an indigenous native, trains himself to see the Bond.

Even though our modern culture, so obsessed with individual things, has taught us to see in a piecemeal and highly focused way, all of us have the ability to recover this capacity to see the subtlety of the relationships that make up our world. The true lesson of the Moken is that the way we see the world is not inherent, but a learned skill. Anna Gislén returned to Surin Island three years later to carry out a second study with the Moken children. This time, however, she created a training program for the European children that was similar to the way Moken teach their own children to see underwater. After just one month of training the European children had learned to notice detail underwater with the same visual acuity as the sea gypsies.[21]

We can change our perception of things by learning to see the holistic connection between things. The indigenous populations of Micronesia have learned to navigate on the open seas for several hundred miles among the scores of tiny Caroline Islands and atolls of the South Pacific without benefit of any instruments or even an ability to write. T. Ward Goodenough, an anthropologist at the University of Pennsylvania, lived with the indigenous sailors for many months and managed to decode an extraordinary oral schematic mapping system that enables them to navigate with great skill.

The sailors create a complex compass system that orients them and the various islands to the rising and setting of all the visible stars in relation to the land. After memorizing all the visible land points, they use various islands as drag points in relation to the stars to keep track of the distance traveled. They also study living "seamarks," such as a species of ray with a red spot behind its eyes that is associated with certain locales, and create schematic mental maps of the islands, imagining them as objects, such as "trigger fish," in their heads. Subtle signs in the sea and the sky enable them to make accurate predictions about the weather. If a certain star appears on the eastern horizon just before dawn, it is likely to bring stormy weather for five days after the next new moon appears in the west at sunset. The shapes of clouds, the color of the sky at sunrise and sunset, the ocean swells and their direction against the stars, even the shape of the waves guide a seasoned navigator along the direction of the current.[22] An indigenous sailor's only aide-mémoire is a tonal chant, with which he constantly reminds himself of the patterns of these relationships.

In order to notice all the relationships between things, the South Pacific sailor breaks down what he sees into manageable chunks and memorizes them in relation to each other. Temple Grandin, an animal trainer and author with autism, believes that indigenous people like these sailors process thoughts in the same way autistic savants do.[23] A person with autism perceives his world in separate bits of incredible detail, observing hidden connections, so that the world becomes a unity of interlocking parts.

Most of us "neurotypicals" process specifics—say, all the sensory data about a building—into a generalized whole, and are, as Grandin puts it, "lumpers." We notice a portion of something and "fill in" the conceptual details to produce a whole. Hence when we see a familiar scene or we're concentrating on something, we become blind to the details of it.

Like the Moken, people with autism perceive their world differently—up close and in great detail. They don't create a unified concept of an object, but perceive pieces of information; they are, as Grandin puts it, "splitters." Rather than seeing an entire object, people with autism see what Grandin terms a "slide show" of the object, with "privileged access to lower levels of raw information."[24]

Like people with autism, animals attend to each tiny, separate piece of sound, sight, and smell. They do not process these raw data into a whole. This type of hyperspecificity develops extreme perception, which is why some autistic children are able to demonstrate amazing feats of memory or an extraordinary talent for discovering "hidden figures" or patterns, such as shapes, within a complicated picture. Every moment of their lives they are flying like a novice pilot.

To begin to see the true connections that exist in the space between things, we also have to learn to fly as a novice does. This requires, in a sense, turning down our cognitive processes and regarding the world simply with our senses. The effectiveness of this has been demonstrated in the psychology laboratory of Allan Snyder at the University of Sydney. In previous research he showed that people who naturally have smaller frontal lobes, as animals do, or a frontal lobe that has been damaged develop hyperspecific perception and greater conscious access to the raw data held in the brain. When he applied magnetic fields to the frontal lobes of normal subjects, they began to make far more detailed drawings than they had before the experiment and were also better at proofreading than before—as if they were suddenly more aware of the detailed parts and the periphery than the summarized whole.[25]

To learn to see the connections between things we need to develop techniques that turn off our overly analytical neocortex and bolster our own innate ability to become aware of this raw data flow. One of the chief practices that creates "lumpers" is thinking in words. Studies show that human language often suppresses visual memory, causing "verbal overshadowing." Native cultures may see more than we do because they often gather information and process it in a manner without language.

Although we may not be able to eliminate our well-ingrained thought processes, we can learn to notice more by acting as though we are experiencing everything we see and do in our lives for the very first time. Ironically, to see the connections between things we have to notice more detail, and the best way to do this is to act like a dog about to go on a walk.

Our dog Ollie is a small, tricolored Cavalier King Charles spaniel, and, characteristic of the breed that was bred by royal decree, he was born with a peculiar sense of regal entitlement and a permanent look of disdain. Ollie belongs in a *Peanuts* cartoon—the curmudgeonly dog whose thought balloon, like Snoopy's, continuously registers his exasperation at his clueless owners. He never comes to the door to greet us. He almost never wants to play. He refuses to eat, except at inconvenient times. On those few occasions we get a lie-in, he pounds on the kitchen door, demanding entry to the rest of the house. Almost all of his time at home he is to be found slumped at the bottom step of our stairs, refusing to move, even when his name is called. Nevertheless all we need to do is to walk toward the drawer where Ollie's leash is kept and he will spring to life, shooting three feet in the air. The very thought of going on a walk suddenly animates him with a joy almost beyond description. It isn't even the actual walk that's good. It's the sheer anticipation. It's so good that he can taste it.

According to Jaak Panksepp, a psychobiologist at Bowling Green State University, Ollie's sudden animation is related to the "seeking" mode of his brain. Panksepp has identified five core primary emotions that humans share with many members of the animal kingdom, including seeking, or curiosity.[26] In seeking mode animals search out things or investigate their environment. This impulse is powered by all the basic needs—for food, water, shelter, sex—an animal needs to stay alive. Nevertheless the most emotionally arresting part of seeking has nothing to do with the goal itself, but the journey along the way. The seeking circuits are fully engaged in animals and human beings whenever we anticipate things, or get intensely engaged in an activity, or become insatiably curious about something new.

As Panksepp was surprised to discover, when animals or humans are curious, the "feel-good" neurotransmitter dopamine is produced. Animal research has also found that seeking sends animals into the equivalent of a meditative state.[27] We are hardwired to feel good when we are sniffing out something new, but only as long as we're on the lookout. The seeking portion of the brain stops firing once the animal has found what it is looking for. Animals have been made to enjoy the foraging and the hunt so that they will carry on doing so until they get what they need to survive. They find it physically pleasurable to be curious, so they persist in looking, even when the going gets tough.

On the whole, wild animals have more of a seeking-circuit activation than ordinary domesticated animals. This is likely due to the fact that wild animals have to keep their wits about them and maintain a high level of curiosity for hunting and foraging to keep themselves alive, whereas domestic animals like Ollie are limited to digging and foraging just for the fun of it. Nevertheless even when survival is not at stake animals of any sort maintain a fascination with the new.

Our own hunting and gathering instincts are no longer important for survival, but we have retained our love of the hunt, whether we're investigating a mystery, having a good snoop around a shop, studying new ideas or projects, or even problem solving. This impulse may even be behind our obsession with detective yarns and whodunits. Indeed, a sense of curiosity may be essential to a long and healthy life. In one study of men and women over seventy a sense of curiosity was the most significant determinant of survival, irrespective of diet and lifestyle.[28] A perpetual fascination with the new appears to be a mainstay of healthy old age.[29]

In the act of seeking, we're firing on all cylinders; it is the moment when we are fully present and engaged in life.[30] In fact the higher our level of dopamine or other neurotransmitters, the lower our level of latent inhibition. This suggests that when you are in a state of intense curiosity about something, you see into the space between things. That appears to be the way we're supposed to see, because nature has designed us to feel so good when we do. Seeking brings out our natural tendency to stay awake and see the whole.

One of Panksepp's important findings is the location of the brain's seeking mode within the hypothalamus. The hypothalamus has been known as "the brain of the brain," the central hub of the body's autonomic regulatory activities, but also the center of the "mind-body" connection, helping to integrate feelings into a cohesive response. It works in harmony with the pineal gland and is part of the limbic system, and it may be involved in what we tend to refer to as our "gut feeling." This suggests that seeking does not entail operating on just one level, but also using emotions and intuition. When we're firing on all fronts we're observing the world from many levels.

Many of the emotions that form our gut instincts could result from two types of information flow, what Joseph LeDoux, a neuroscientist at New York University, refers to as unconscious "low-road" information to the amygdala and conscious "high-road" cognitive information to the neocortex. According to LeDoux, fear travels to the conscious part of the brain slowly but to the unconscious part (the more primitive amygdala) in a few milliseconds. This has an obvious survival advantage. You run away from something potentially dangerous before you take the time to work out in your frontal lobes whether it actually poses a threat.

Another way to learn to see the Bond is to practice the ancient art of mindfulness, espoused as long ago as 500 BC by Shakyamuni Buddha, the founder of Buddhism. In this discipline you maintain a moment-to-moment awareness of what is happening internally and externally, without coloring your interpretation with your emotions or distracting yourself with other thoughts.

Mindfulness requires that you maintain your concentration in the present. With practice you become able to silence the constant inner chatter of your mind and meditate on the focus of your concentration, no matter how mundane, whether eating your cornflakes, smelling a flower or bending down to tie your shoe.

Mindfulness has been termed an ancient method of sensitivity training to your own life. You learn to listen with full attention, see what is really there, engage all five senses, and remove attitudes, judgments, and concepts that color your perception of everyday experience. When you engage in mindfulness you discover what you are thinking and feeling without being wrapped up in and enslaved by your thoughts.

The ancient Buddhist texts claimed that daily practice of such attentiveness will significantly change perception, and science bears this out in research demonstrating that our brain operates at different frequencies during mindfulness. In one study EEG tracings taken while participants practiced mindfulness meditation showed that mindfulness produces different electrical frequencies in the brain, increasing certain bandwidths.[31]

Another study tested the visual sensitivity of participants before and after a three-month retreat where they practiced mindfulness meditation for most of their waking hours; staff members who did not meditate were used as a control. The researchers assessed whether the participants could detect the duration of light flashes and the interval between successive flashes. To those without the benefit of mindful practice, the flashes would appear as one unbroken light.

After their retreat the mindfulness practitioners could detect the shorter single-light flashes and could also correctly differentiate between successive flashes. These results demonstrate that mindfulness enables its practitioners to become aware of pre-attentive processes involved in visual acuity and undergo a long-term increase in sensitivity of perception.[32] They begin to see like a Moken swimmer.

Over time mindfulness meditation causes enduring changes in mental function, so that you *can maintain simple awareness of the whole* of the stream of stimuli in front of you, without judgment or a narrowing of focus. You remain *alert to nuance and detail*, particularly the unexpected and the seemingly contradictory, and are less likely to zero in on a single image or idea.[33]

After just three months of intensive mindfulness participants in another study increased their ability to take in all of the sensations and events in front of them. Instead of focusing on particular visual targets, they perceived a rapid stream of minor distractions that they would ordinarily ignore.[34] Mindfulness also reduces the endless judgments (Do I like this or not? Is this good or bad?) that usually accompany perception.[35] When you become fully present, you commit to seeing the whole of your life, not simply what you prefer to see.

With practice mindfulness also opens you to intuitive forms of information, the communication between people that lies beyond language and the senses. You become sensitized to the emotional undercurrents of relationships and increase your own capacity for empathy.[36]

Don Beck is convinced that if he had been alive in 1860 and had a conversation with Abraham Lincoln, he might have been able to prevent the Civil War. A former professor of social psychology at the University of North Texas, Beck, a florid seventy-five-year-old Texan, is best known for a system he developed called Spiral Dynamics, which identifies the fine gradations of belief systems and their level of complexity in any given society. He considers his work a continuation of his doctoral dissertation, which examined the polarization of Americans just prior to the Civil War. Beck discovered no fewer than eight political positions on slavery, from those in favor of unpaid servitude to those desiring full abolition. When the moderate position disappeared from both sides, he says, the country polarized and the war began. "If we had done certain things in 1860, we would have ended slavery and we wouldn't have lost 700,000 people," says Beck. "And we wouldn't still be fighting the Civil War."[37]

As a political consultant on resolving societal conflict, Beck calls himself a human "heat-seeking missile," drawn to the world's hot spots: South Africa, Palestine, Afghanistan, Israel. His current work attempts to break up the thinking that fuels us-versus-them prejudice by showing people on one side that those on the other side are not all the same. In Beck's experience, what often polarizes people or pulls societies apart is a simple lack of appreciation of the spectrum of different beliefs that exist in cultures outside their own. "We don't have the language of difference, so we tend to stereotype," he says.

During his sixty-three trips to South Africa in the 1980s Beck became known as a bridge builder between the country's black and white populations; as a consequence he played a behind-the-scenes role in helping to smooth the country's transition from apartheid to democracy. In his dealings with the business community he began to realize that many of the pro-apartheid Afrikaners, the dominant white group, were unable to differentiate between various black tribes, while members of the African National Congress, the party led by Nelson Mandela, also had difficulty distinguishing between different types of Afrikaners. Beck began delivering presentations all over South Africa to educate whites and blacks in the fine distinctions between the many different black and white groups. "I was able to break up," he says, "the definitional systems that fueled prejudice."

You can develop this kind of "aerial vision" in your own life by fine-tuning your ability to notice the details about people and cultures that are different from yours, which eventually helps you to refrain from thinking that traps you into an "us" and "them" mind-set. You learn to question your automatic assumptions about the neighbors you don't know, the acquaintances of a different ethnic or religious persuasion, the countries and people beyond your borders.

Aerial vision also enables you to hold opposing ideas in your head without being judgmental, because you recognize the whole that ties everything together. You detach from your own vantage point and your own prejudices in order *to entertain several viewpoints and stop taking your own side*. As a leadership facilitator with the U.S. Congress and the United Nations, Mark Gerzon once worked with the New Israel Fund, an organization whose membership, comprising both Israelis and Palestinians, raises money for a number of joint causes. When he asked the group's board how they were able to work together so effectively, one member replied, "We are able to live with paradoxes."[38]

To illustrate their point, one Palestinian and one Israeli each wrote his own abbreviated version of the establishment of the State of Israel. The Palestinian narrative, called "The Disaster," referred to the establishment of Israel as "a great tragedy": "Entire villages were destroyed and land and property were confiscated while hundreds of thousands of Palestinians became refugees. Overnight, the Palestinians became a minority in their own homeland."

The Israeli's narrative, on the other hand, titled "Independence," marked the fulfillment of the Jews' "historical right" to establish a state in the Promised Land. Nevertheless, despite the UN's partition plan, which declared Israel a state, from "the dawn of its foundation," the Jews were beleaguered from all sides. "The Arab nations attacked the Jewish nation," but "the Jews bravely expelled the British mandate, and since then, the Jewish nation has been fighting for its mere existence. . . ."

The tragedy—the reason for the enduring conflict—is that *both* stories are essentially true.[39] When you notice the whole, you allow for and respect more than one version of reality.

Aerial vision also allows us to transcend our tendency to always look to our own vantage point for solutions to problems. Well-meaning U.S. troops caused enormous resentment in Iraq, for instance, when they demolished several beloved and well-used soccer fields in Baghdad in order to build the $1.5 million Tigris River Park as a "good-will gift."[40]

Recently Don Beck attended a large investment conference in Bethlehem, where a number of Western multinationals were recommending high-tech investment in Palestine. To their astonishment Beck insisted that they invest in a cement factory. The potential investors were alarmed by the idea of investing in Industrial Age technology until Beck convinced them to look through Palestinian eyes. The refugees were living without permanent homes; what they needed most was not computers or mobile phones, but predictable blue-collar work, an industry with a constant market, and a source of building materials inside their own borders.

Once we begin to see the whole we can see past our own assumptions and beyond differences to our common humanity—and to the space that binds us all together.

CHAPTER 10

SURRENDERING
TO WHOLENESS

Orland Bishop appears to be on a fool's errand, instructing young
black gang members in the art of communication. Bishop's chosen
patch to spread the word is Watts, a district in south Los Angeles
considered so devoid of the possibility of redemption from decades
of entrenched poverty and violence that it is now chiefly known for a
string of negative achievement. In the mid-1990s the street gangs of
Watts formed the epicenter of America's crack cocaine business. Such
was the rivalry between the main gangs, the Crips and the Bloods, that
the wars between them and their offshoots claimed five times as many
lives as did all the years of the troubles in Northern Ireland. The white
establishment's solution, the creation of the special CRASH division
(Community Resources against Street Hoodlums) in the Los Angeles
Police Department, resulted in the largest internal affairs investigation
of what turned out to be the largest incidence of police misconduct
in America's history—for unprovoked shootings and beatings, framing
suspects, planting evidence, and the department's own share of drug
dealing and bank robbery.

The very name of the district has become synonymous with the most
extreme face of racial discrimination in America. During the riots of
1965, set off by the arrest of a black youth and his family on a trumped-
up drunk-driving charge, black residents burned and looted nearly
a thousand mostly white-owned businesses. Some 15,000 troops of
National Guardsmen and Armored Calvary, more than had ever been
deployed on the nation's own soil, were called in, ostensibly to prevent
Watts from burning itself to the ground. "Monkeys in a zoo," is how

the LAPD's police chief, William Parker, publicly summed up the situation, sparking off another round of looting and arson contained only when Guardsmen cordoned off all of Watts from the rest of Los Angeles like an epidemic requiring quarantine.

Thirty years later, in 1992, Watts rose again as the symbol of the uneven hand of American justice after a tourist's video captured policemen brutally beating a black motorist, Rodney King, and the officers involved were acquitted of all charges. The court decision set off a six-day riot of arson, assault, and murder, leaving fifty-three dead, thousands injured, and more than one billion dollars in damages.

Although the murder rate in Watts has decreased since the 1990s, most attempts at urban renewal tend to wither on the vine. The riot after the officers' acquittal, this time targeting the community's upwardly mobile Hispanics and Asians, occurred the day after the four major black gangs had signed a peace treaty. The city's single source of public recreation, the 109th Street Pool, required constant police reinforcement after youths attacked its manager in 2005. In the ongoing dialogue between the races in America, Watts is the place where no one is doing much listening.

Nevertheless it is here, in this No Man's Land, that we can learn how to recover the Bond in our relationships. Orland Bishop's work teaches rival black gang members how to relate to each other. If he were to offer you and me similar instruction, it would differ only in degree, not kind. Shooting each other is simply the most extreme version of the kind of relationships you and I engage in when we act against nature and communicate from a selfish, competitive point of view.

Bishop, an angular forty-four-year-old with a graceful bearing and a singular eloquence, embodies his own interesting contradiction. He was born in South America but came of age in Brooklyn, and his work developed in the tension between these two worlds. He grew up in Guyana with a profound sense of connection with everything around him: his family of nine, with whom he shared a two-bedroom house; the natural world, where he spent summers wandering through the rain forest; and even his fledging country. In 1966 Guyana's newly won independence from Britain sparked a new collective vision. Bishop and his friends savored their long walks and deep conversations. Everything was there to be shared.

Bishop was abruptly severed from this connection in 1982, when his parents moved the family to the United States to live with an aunt in Brooklyn. Even though Guyana's British system of study placed Bishop two years ahead of the New York curriculum, he was put back two years because he had come from a Third World country. To his mind Brooklyn, at its peak of drug-fueled crime at the time, was the uncivilized place, not tiny countries in the Caribbean. Upon graduation he went to college to study medicine, instinctively understanding that the path to his own connection to his new surroundings lay in healing the damage he witnessed around him.

During his medical studies a talented artist friend came to him seeking help for his drug addiction. When his friend's problem turned out to be AIDS, Bishop became his primary caregiver.

"He invited me into the conversation that I am still in," says Bishop. "A few days before he died, he turned to me with the most compelling force of human will that I have ever witnessed, and said, 'You better do what you are here to do. If you don't do it, you will regret it.'"[1] Bishop realized that the healing he was destined for lay beyond the physical body, in the disconnection within American society. He quit medical school and began the ShadeTree Foundation to tutor gang members and at-risk black youth in the possibility of having relationships without aggression or violence.

Bishop believes that a gang is simply a manifestation of the thwarted human need to belong. "They have an instinct toward oneness, which is why they form gangs," says Nelsa Libertad Curbelo Cora, a peace worker with young gang members in Guayaguil, Ecuador's version of Watts. Bishop's work is all about teaching his young gang members to move beyond "I" and "you," or more commonly "us" and "them." It is radical precisely because he recognizes that to recover the Bond we must dampen down the "I" in our own story.

In his seminal work *I and Thou,* Martin Buber claims that, in the main, we relate to other people as "I-It," as objects utterly separate from—and hence subordinate to—ourselves.[2] That is largely because, in any situation and in any relationship, we consider "I" to be separate and primary. As Richard Nisbett discovered in his studies on the differences in the way Easterners and Westerners think, our individual relationships have a lot to do with how we see ourselves in relation to rest of the world. When asked to describe themselves, North Americans and Europeans tend to stress their individual personality traits, exaggerate their uniqueness, and focus on what they regard as most distinctive about themselves and their possessions, whereas East Asians stress their relationship to all social groups.[3] Attending to the separate object entails seeing the world—categorizing, applying rules, and working out causal relationships—entirely from the vantage point of the object. As Westerners, when we walk down a suburban street, the houses are naturally to *our* left, the cars and the street to *our* right. Everything is oriented in relationship to us, as though we are the sun and the rest of the universe our planets.

In one of Nisbett's studies he gave American and Korean students the choice of different colored pens to keep as a gift. The Americans chose the rarest colors, whereas the Koreans chose the most common. One sought to be the most distinctive person in the room; the other wanted simply to fit in.[4] In the Western world our impulse to be separate and unique in a relationship often translates into a need for power. In most of our encounters we are compelled to prove that it is we who have the flashiest pen.

But even when we seek to connect rather than to compete we usually continue to focus on the "I" of the story. If I were to ask you to describe the first meeting of most of your friends, you would probably recount how you first cast about for points of mutual contact, evidence that you share the same economic level, spiritual beliefs, hobbies, family structure, or personal tastes. Most likely you have chosen to connect exclusively with people who share something of you in them. We think of this superficial connection as providing us with a sense of shared identity.[5] We like people who are just like us, who share our values, our attitudes, our personality, and even our emotional disposition,[6] and we tend to conflict most with people who are not like us. All of the groups we join, from the Rotary Club to the Parents and Teachers Association, are based on a shared passion, whether a community, game, God, or children. Our idea of connection is constantly seeking sameness. What this means of course is that the ultimate yardstick by which we gauge everyone else is ourselves.

This tendency to cluster with people who are most like us serves only to divide us from others by reinforcing our individuality, our sense that our way is the best. We always look to re-create ourselves in another, which has as its basis a desire to reinforce that we are *right*. This also runs counter to our deeper biological experience of relationships. As we have seen, the moment we leave our own solitary company, we temporarily depart from ourselves and automatically enter the other—no matter how different—and they also enter us. We are affected by the feelings, actions, and thoughts of all others, and they in turn are affected by us, even if we have nothing in common. For better or worse, whether we wish to or not, we are merging into everyone with whom we come into contact. The Native American Lakota tribe refers to relationship as *tiyospaye*, which roughly translates as "where we live is apart, but not separated completely." No matter how distant or disagreeable, in every interaction, at some level, we feel the Bond.

We are constantly attuned to all other human beings, and can become conscious of this if we pause long enough to listen. But that requires moving beyond a simple clustering of like with like to find the deeper connection that is always present in any relationship. When we feel the Bond, we reduce our sense of subject and object and train ourselves to see "I" only in relationship to the other. This demands that we shift our focus from the smallness of the self to the expansiveness of the space in between.

After his friend with AIDS died, Bishop traveled extensively through western and southern Africa, studying with Zulu tribal elders, including Credo Mutwa, the last of the Sanusi elders. When he returned to America, Bishop brought back with him a means of relating embodied in the Zulu greeting *Sawubona*. Although usually translated as "I see you" (and made famous by the N'avi in the film *Avatar*), *Sawubona* literally means "*we* see you," and the correct response is *Yabo sawubona*, "Yes, *we* see you too."

"It's not a single 'I' person," says Bishop. In African culture a person is thought to be connected to all living things and the whole of consciousness, past and present, and so relating is never thought to be a solitary act. The notional plural is also important because, as Bishop puts it, "seeing is a dialogue," an act of bearing witness and an obligation to participate. You bear witness to something that includes your own presence and that of the other. What Bishop is driving at is the essence of the Bond: during every moment we relate to another human being, we make, in a sense, a contract—to enter the space in between, the space of pure connection—regardless of any superficial differences.

"It's an invitation for us to participate in each other's life," he says, "to investigate our mutual potential for life. It also obligates one to support the other—to give to each other what is needed for that moment of life to be enhanced." This sounds like something akin to the African concept of *ubuntu,* which in its literal meaning, "I am because you are," suggests that as cocreators of each other, both observer and observed, we have a commitment to provide to the other what is needed at that moment, whether food, water, or the deepest level of support.

Engaging with another in any way obligates us to share the moment to the best of our ability, to offer exactly what is required to help the other to flourish. Feeling the Bond changes the nature of the transaction between you and someone else from a selfish aim to a broader focus, whose sole purpose is the connection between the two of you. Bishop tells his young people to regard each encounter with someone else as a personal Challenge: "How do I have to be for that person to be free—to be who they are? What forms of cognition do I have to engage in to allow them to feel free enough in this relationship?" In this definition of relationship the sense of "I" and "you" grows to something much larger: the moment of the Bond when difference doesn't matter. Offering yourself as a vehicle of service to the pure experience of connection can occur with anyone, simply through the fact that, as two human beings, you are both taking in breath. "Through Sawubona," says Bishop, "we are capable of experiencing the quality of another without judgment or prejudice shaped by our thoughts. Sawubona is an openness to the highest good in a person."

Once we view ourselves as part of a larger whole we begin to act differently toward each other. By removing a self-serving aim from the relationship, we stop fighting nature and surrender to our natural impulse toward holism. We can easily embrace difference within that larger definition of connection.

Bishop talks a good deal about space: the physical space we inhabit, the psychic space we offer someone when we relate to him properly; the space where we meet on some deep level. He regards "space" as synonymous with "sanctuary," and "sanctuary" as synonymous with a perfect kind of connection. *Sawubona,* he says, "brings us into a relational space—a shared space—with another and within the community."

Because most LA gangs are fighting over territory, their common space is Watts, which has an economic context for them, usually in the form of drug sales. Bishop neutralizes this context of territory by removing the possibility of exclusive ownership, so that territory can be viewed as a sanctuary rather than a source of income, where it is possible to share both physical space and identity. Relating should also offer this temporary psychic sanctuary, so that both parties feel heard and understood. The focus shifts from an exchange of services or a search for sameness to a sharing of the deepest aspects of each other. Bishop invites the young men to engage in *indaba,* which is loosely translated as "deep talk," moving past superficiality to the deepest truth of who you are and what you dream of.

"For people to be in agreement they must be in a deeper perception of the powers of their own humanity," he says. When you share this deeply, as he suggests, you surrender to your natural impulse to merge together and you find the common ground of the space between you—the place of your common humanity. "Shared meaning," says Bishop, "allows for different perceptions—or realities—to exist together."

Bishop holds five-day residential retreats in northern California, deliberately moving young members of both gangs away from familiar territory and using ceremonies and rituals to shake them out of their ordinary patterns of behavior. During unstructured discussions he asks the group provocative questions—"When did you begin to feel exiled from any chance to be what you really wanted to be part of?"—to spur the young men to share their personal histories. Most are shocking, and the young people still very raw from them. "Often a wound drives a young person to the gang experience," says Bishop. "They think they need an enemy to be themselves."

Bishop coaches the young men in the art of *speaking and listening deeply and from the heart* without being critical or judgmental. During this type of deep sharing the pull of wholeness builds trust and loosens their attachments to entrenched positions. The very intensity of the experience lends itself to the establishment of new alliances and a larger vision for the future. "They begin to understand that if they unite in creativity they expand," he says.

One deep connection such as this can be remarkably healing. "Once you have this relationship dedicated to that greater good," says Bishop, "it can hold the space for it to just keep forming, no matter what the current reality is."

In Watts the Crips and the Bloods enjoyed a twelve-year truce, presently being renegotiated. For the past eight years Bishop has worked on helping them craft a future that they both can share. Within its large network ShadeTree includes a number of young men who were in rival gangs now working together. Similarly, in Guayaguil, the peace worker Nelsa Cora has taught young people in street gangs to transform this need for connection into "the power of service, life, and love" for a struggling community. Gang members have learned to channel their impulses for creativity and need for recognition away from violence and into small businesses: printing businesses, music studios, pizzerias. The barrio is now known as Barrio de Paz, Peace Town. Surrendering to the space in between is so powerful that even the most ruthless of gang members have laid down their guns.

Indaba resembles the art of dialogue, first proposed by David Bohm, a communication skill in which a group explores feelings and ideas in an unstructured way to create greater understanding, deeper connection, and a new synergy of ideas. Bohm, a British physicist who believed in the unseen unity of all matter, was convinced that our thoughts were also part of a unified field and that the crises facing mankind had to do with a "pervasive incoherence" in the way modern people think.

What we consider our private thoughts are in fact a collective phenomenon—the result of a mass convergence of cultural influences. Our version of reality is simply a construct, colored by our own concepts and memories, which themselves are influenced by language, culture, and our own histories. Nevertheless, wrote Bohm, each of us believes that the way we interpret the world is "the only sensible way in which it can be interpreted."[7] Consequently when attempting to talk about subjects that matter most to us, we speak from our own version of the truth and invariably end up disagreeing with anyone whose version slightly differs from ours.

Bohm proposed a method of discourse that would slow down the process of thoughts so that individual and collective presuppositions could be explored and the ideas, beliefs, and feelings behind them exposed. By helping individuals to understand the processes that interfere with real communication, this kind of exchange would establish a common version of reality. Bohm likened dialogue to a river flowing around and through the participants, reinforcing the Bond between them and creating a "coherent culture of shared meaning."[8]

The rules of dialogue are simple. The parties involved agree that the purpose of the conversation is *not to reach a decision or to have a debate*. All agree to *take turns* speaking and not to monopolize the conversation. The participants also agree to be *alert to their own reactions* as they are taking place. Dialogue makes use of the power of *deep truth-telling*; individuals agree to be honest and transparent about the areas of greatest importance to themselves, no matter how controversial or contentious their position, and never to use their views to make another person wrong.

All agree to *be fully present and listen respectfully with both heart and mind, without judging each other*. Each member tries to build on the ideas of others in the conversation and engage differences in creative ways in order to produce greater shared understanding, connection, and possibility. You begin to commune in ways that enable you to connect directly from your soul to the soul of the others in the group in order to find the common humanity and connection beneath superficial differences. In the free-flowing interplay of ideas a "pool of common meaning" evolves. "A new kind of mind," Bohm noted, "comes into being."[9]

In December 1989 Laura Chasin, a family therapist in Cambridge, Massachusetts, was watching a rancorous debate on abortion between pro-choice and pro-life representatives. The argument reminded her of some of the same kind of behavior patterns she regularly dealt with in her practice with dysfunctional families. She wondered if some of the techniques that proved effective in therapy could also be applied to people whose political or social views were polarized.

Chasin created the Public Conversation Project by enlisting women friends and acquaintances on both sides of the abortion issue to deepen their understanding of each other by changing the way they communicate. She began holding meetings over buffet dinners, where the women could get to know each other before they disclosed on which side of the fence they stood.

In subsequent meetings the women sat in a circle and took turns in a dialogue, revealing their personal stories about abortion, the events in their lives that helped to shape their beliefs, the aspects of the issue they still wrestled with. In total Chasin hosted eighteen sessions with more than a hundred different women.

Then, on December 30, 1994, when the pro-life advocate John Salvi shot dead two and wounded five others at the Brookline, Massachusetts, Planned Parenthood and nearby Preterm Health Services, six leading figures from the state's pro-life and pro-choice movements, including the director of the Pro-life Office of the Archdiocese of Boston and Nicki Nichols Gamble, the director of the Planned Parenthood League of Massachusetts, decided that it was vital that the two sides continue the dialogue. The six women carried on meeting in secret for nearly six years.

Over time they learned to stop using inflammatory language such as "murder" and learned to "speak in love, speak in respect, and speak in peace," no matter how wide their differences.[10] At a service to honor the memory of the two who had been killed in the Salvi shootings, Gamble expressed gratitude for the "prayers of those who agree with us and the prayers of those who disagree." Both sides joined forces to announce that the head of ProLife Virginia, who reportedly had sanctioned Salvi's action as a "righteous deed," was not welcome in Massachusetts. In addition, the pro-life leaders created a hotline system to alert the pro-choice leaders whenever they thought there was the possibility of violence or physical danger.

At the end of the six years the group held a press conference, at which members of the press wanted to know who had "won" the debate. Each of the six announced that the process of the dialogue had helped them to become firmer in their own views about abortion.

"So, it was a failure then?" asked a reporter.

"Oh, no," replied one of the women. Although they had struggled with profound philosophical differences over the years, they had found the Bond between them and discovered how to treat each other with dignity and respect. *Now, you see, we party together. We watch each other's children. We love each other.*[11]

Bishop and the Cambridge dialogues both focus on communication that *reveals the deeper narrative* of each person's life—how we come to believe what we believe and who we really are—and the *connection that always exists at that deeper level of being.* The Cambridge women deliberately avoided sitting across the table in a confrontational manner, as is usually the case in negotiating or decision making, but sat companionably side by side, in a circle. They sought *higher ground*—creative ways in which to work together—to offer better sex education to teenagers, greater help for pregnant teenagers, and improved adoption programs. However, the most important aspect of the dialogue is *a surrender to the pull of wholeness,* seeking out the "pool of common meaning." By walking into the space between, you discover the common ground that is always there, even when worldviews collide.

By opening ourselves to our truest nature, which always seeks wholeness, we allow the possibility of pure and immediate resonance with the other within the space of our common humanity. This is not particularly difficult if we reach into our memories. In fact most people first experience this pure connection as parents, or even at the moment of their own birth.

I was certain that I was going to be a bad mother. I had waited to get pregnant until well into my thirties and spent a good deal of my pregnancy convinced that I, so invested in my career, lacked the equipment necessary for the selfless role required for mothering. I approached pregnancy like a journalistic assignment, attending the prenatal classes of a natural birth pioneer, reading books, twisting my body into all sorts of primal positions, even carrying out trial runs by holding the babies of friends.

So concerned was I about my appropriateness for the job that I spent the final days of my pregnancy trailing after my friends who had already given birth, jotting down notes in my reporter's notebook, in the hope that something of their effortless and skilled attendance to the all-consuming needs of a helpless being might rub off on me.

The central revelation of our elder daughter's birth, aside from the miracle of it—the extraordinary primal experience of being a vehicle for someone else's entry into life—was how relatively automatic the caretaking part of it was. Although the needs of a newborn are overwhelmingly time-consuming, nothing about the job is particularly difficult. To my astonishment, from the first day of Caitlin's life I seemed to be able to immediately intuit what her cries meant and to make them stop.

"Oh," I'd say casually at the first sign of a quivering lip, "she needs to be changed." Or "She's hungry." Or "A first tooth is coming in."

When Caitlin was fussy, no amount of cajoling, cooing, rocking, tuneful singing, or pulling faces by my husband could match simply handing her over to me, at which point her crying would abruptly stop. I often felt like a sorceress whose very being acted as an incantation to keep my child permanently under my spell. What I hadn't realized then was that I was, in a sense, a metronome for her brain waves, setting the tempo for the resonance between us. Our brain waves had coordinated into a single undulation. She stopped crying whenever she was handed to me because I had reestablished our ongoing resonance, and each time we met in the space between us.

Opening yourself up to a pure connection with someone else, as occurs with a mother and child, creates a neural resonance effect between you. This kind of deep resonance can occur with anyone, not simply our children. We are all so attuned to the emotional states of others that the brain is constantly in readiness to resonate with them. The limbic system, which sits between the brain stem, which is responsible for our autonomic nervous system, and the neocortex, the modern problem-solving and conceptual brain, is known as our "feeling brain." The limbic system is thought to pick up on and interpret the emotional complexion of experience, including that of others. All mammals possess a limbic brain, and therefore are capable of what UCSF psychiatrists Thomas Lewis, Fair Amini, and Richard Lannon have termed "limbic resonance," enabling a living thing to tune into another's inner emotional state. Limbic resonance describes a situation beyond simple mirror mimicry; it is when two people find themselves completely in synch, momentarily melding into a single organism.

One of the most memorable descriptions of this process comes from the naturalist Annie Dillard, who came upon a weasel near her home in Tinker Creek:

> The weasel was stunned into stillness as he was emerging from beneath an enormous shaggy wild rose bush four feet away. I was stunned into

stillness twisted backward on the tree trunk. Our eyes locked, and someone threw away the key.

Our look was as if two lovers, or deadly enemies, met unexpectedly
on an overgrown path when each had been thinking of something else: a
clearing blow to the gut. It was also a bright blow to the brain, or a sudden beating of brains, with all the charge and intimate grate of rubbed
balloons. It emptied our lungs. It felled the forest, moved the fields, and
drained the pond; the world dismantled and tumbled into that black
hole of eyes. If you and I looked at each other that way, our skulls would
split and drop to our shoulders.[12]

Dillard is depicting the shock of recognition that occurs when we
feel the Bond, which is all the more astonishing for her because the
deep connection occurs with a wild animal. But her description extends
to the Bond we create with another person whenever we move beyond
selfish aims; within seconds, as the San Francisco psychiatrists put it,
"two nervous systems achieve a palpable and intimate apposition."[13]
Although we are always mimicking each other involuntarily, we are
most likely to get on each other's brain waves when we establish a deep
connection of the kind suggested by Orland Bishop.

It is not difficult to stoke this natural impulse to seek shared connection. As noted earlier, it has been demonstrated scientifically in the
extensive studies of what is usually referred to as "direct mental influence on living systems," in which a "sender" who has been stimulated
by something—a flashing light or an electric shock—attempts to transmit a mental image of the stimulus to his partner. As he does so, many
of the physiological systems of the pair begin to synchronize.

Research has shown that certain conditions can amplify this effect.
At the University of Mexico the neurophysiologist Jacobo Grinberg-
Zylberbaum found that his senders and receivers were most likely to
mirror brain patterns evoked by a flickering light if they had connected beforehand, by spending twenty minutes with each other in
meditative silence.[14] Other methods of establishing deep connection
have included exchanging an object or photograph, holding each other's hands, meditating together, or simply being designated as partners.
Researchers at the University of Washington demonstrated that even
strangers develop strong brain-wave resonance if they have been paired

off. Sending healing creates brain-wave entrainment, as does an act of touching or mentally focusing on the heart.[15] Strong motivation, experience in meditation, and shared beliefs all create a situation in which two people rapidly get into bodily synch.

The most important factor of all may be cultivating a nonspecific altruistic love. In the IONS Love Study, described in chapter 4, in which one member of a couple sent healing thoughts and intentions to his partner with cancer, the most crucial component appeared to be instruction in pure compassion. The IONS researchers trained senders in a technique based largely on the Buddhist idea of "giving and receiving," a practice that teaches people to develop a deep empathy for others, to understand someone else's suffering without being burdened by it, and to transform this understanding through the process of sending healing.

Called Tonglen, this compassionate meditation technique is progressive, beginning on a small scale of appreciation of others and opening out until it encompasses love for all living beings. The meditation starts with the thought "I appreciate the kindnesses and love of all living creatures." You then think about those closest to you and pray, "May they be well and free from suffering." Then you move on to bless your good friends, your acquaintances, and finally even your enemies.

Richard Davidson, a psychologist at the University of Wisconsin, spends his life attempting to locate where a particular emotion is triggered in the brain. Davidson was particularly intrigued to discover that people with negative emotions have persistent activity in regions of the right prefrontal cortex.[16] He wondered whether the opposite was also the case—that positive emotion led to more activity in the left frontal lobes. His brain research came to the attention of the Dalai Lama, who decided to lend the University of Wisconsin a group of senior Tibetan monks who were highly skilled in meditation. When Davidson tested one of the monks, he discovered more activity in the left frontal lobes than he had ever recorded. From his neuroscientist's perspective, this was the happiest man he had ever seen.

These results also suggested something even more significant to Davidson. Neural registration of emotion appeared to be highly plastic, or mutable, a learned skill that could develop with certain thoughts over time. Davidson went on to conduct extensive experiments with Buddhist monks to determine whether meditation actually conditions the mechanism of the brain to enable a person to become happier and more empathetic.

Davidson and his colleagues recruited other Buddhist meditators highly experienced in a type of practice called "compassionate meditation" (*Nyingmapa* or *Kagyupa* in the Tibetan tradition), as well as a group of volunteers from the university community with no experience in meditation. The novice meditators were given instruction in meditative practices written by Matthieu Ricard, a French-born Buddhist monk with a Ph.D. in molecular genetics, who acts as the main interpreter for the Dalai Lama. For one hour a day for a week the new meditators were to engage in a twenty-minute practice in "pure (non-referential) compassion," an unconditional readiness and availability to help all living beings. To train the participants in this state of mind, Ricard instructed them to think about someone they cared about, such as a parent, sibling, or partner, and to allow their mind to be invaded by a feeling of altruistic love (wishing well-being) or compassion (wishing freedom from suffering) for that person. Ricard predicted that after practicing this exercise for a short time the novice participants would be able to generate a nonspecific feeling of love and compassion toward all living things.

Both experienced and novice meditators were then placed in MRI scanners and told to adopt a state of compassionate meditation or a relaxed state of no emotion, positive or negative, while Davidson played various sounds—a woman in distress, a baby's joyful laugh, or simple background noise—while measuring their brain waves and comparing activity during meditation and the relaxed, neutral state.

As Davidson predicted, brain activity was greatest during times of compassionate meditation while the baby's cry was played. Although it was far greater in the expert meditators than in the novices, even the new meditators showed a greater sense of generalized compassion during meditation than when at rest.[17]

Davidson's study holds many keys for learning how to feel the Bond. Practicing compassionate meditation regularly can sensitize us to others permanently, so that we are available to others in the way that Bishop suggests. Davidson's monks possess a brain circuitry of compassion that is more active than normal, even during rest. As his research of novice meditators demonstrates, this ability can develop fairly quickly. The Buddhist practice of compassionate meditation is very much like our natural impulse to give of ourselves unconditionally.

Most fascinating about Davidson's work is the area of the brain where effects occur: the temporal-parietal junction, in the right hemisphere. The neuroscientist Andrew Newberg of the University of Pennsylvania has discovered that during certain forms of meditation the activity of the parietal lobes is decreased. This part of the brain helps us to generate a three-dimensional image of our bodies and to orient ourselves, and so enables us to work out *self* from *not-self*. Newberg also discovered that a practitioner of compassionate meditation loses the sense of self and other and enters a perception of oneness.[18] This unconditional readiness to give—part of our natural impulse to connect—helps to dissolve individual boundaries, enabling us to step out of our individuality and into the space between.

The pull of wholeness inside each of us can be called upon to restore harmony after deep differences or transgressions. As codirector of the Social Healing Project, James O'Dea, former director of the Washington, D.C., office of Amnesty International, has spent many years smoothing the way for warring sides to reconcile and forgive. For ten years he and Dr. Judith Thompson cohosted "compassion and social healing" dialogues, in which members of widely divided social and political groups—Republican and loyalist Northern Irish, Turkish and Greek Cypriots, Israelis and Palestinians—meet in an attempt to heal shared wounds.

In the dialogues O'Dea and Thompson move the emphasis away from who is right and who is wrong and toward who is wounded and how to heal. The aim is to help each party to recognize the other's pain or shame and, in so doing, to liberate each other from hurt and guilt.

Their method draws upon the work of the theologian Geiko Müller-Fahrenholz and his book *The Art of Forgiveness*.[19] Born in 1940, Müller-Fahrenholz was too young to have any memory of the Third Reich, but like so many postwar Germans he grew up haunted by his country's terrible legacy and so began to consider forgiveness from the perspective of both victim and perpetrator.

Müller-Fahrenholz considers wrongdoing a mutual bondage. Any such act, including the most minor of transgressions, establishes a distorted relationship between two people. The perpetrator has stolen power, and the victim has had impotence thrust upon him. For the victim, hurt is an "impairment of the core of our personhood," he writes.[20]

Forgiveness can never replace justice, but it can move us beyond the simple notion that retribution is enough. In Western culture, which largely uses punishment and imprisonment to deal with transgression, both victim and perpetrator remain in bondage. The victim's dignity and personhood (or goods) are not restored, and the perpetrator never truly comes to grips with what he has done.

An act of forgiveness, on the other hand, as the philosopher Hannah Arendt once wrote, is a "constant mutual release." Both victim and perpetrator learn to recognize each other's pain or shame and mutually liberate each other from hurt and guilt.

Müller-Fahrenholz tells the story of a group of old Germans who had fought in Belorussia as part of Hitler's army during World War II. They decided to return to the country in 1994, fifty years later, in an attempt to make amends for what they'd done as young men. Their visit occurred after the Chernobyl nuclear accident, so they offered to build a home for children affected by the disaster. Toward the end of their stay they visited a war memorial at Chatyn. That evening, full of the memories brought up by the visit, the Germans wanted to share the experience with their Belarusian hosts.

After a round of very personal toasts, one of the Germans, still clearly overcome by his visit to Chatyn, stood up to talk about his own history as a young soldier. He began describing his own suffering in a Russian prison-of-war camp, but abruptly stopped. He excused himself for a moment and then suddenly broke down. He said how deeply sorry he was for what he personally had done to the Russians and also apologized on behalf of his country.

He tried to say that it must never happen again, but his voice again broke, and he had to sit down because he was sobbing so hard. Everyone in the room, even the young people who had no experience of war, were weeping.

After a few minutes a Belarusian woman of similar age stood up, crossed the room, and kissed him.

At the moment of the German's genuine act of confession, the full hurt was acknowledged and the dignity of everyone in the room was restored. For the old woman, forgiveness was sparked by the sudden realization that the pain of others—even the pain of the perpetrator—was also her pain and that of every one of the victims.

This moment of connecting to the other's pain is the transcendent aspect of any relationship, writes Müller-Fahrenholz, offering "a spark of courage to open up, that moment of daring and trusting which causes the heart to jump over the fence."[21] Ultimately it is this sudden merging that tears down "the dividing walls" between us.

Deep truth and candid disclosure interrupt the cascade of denial and, most important, *reconnect the Bond by reestablishing the balance in the relationship,* far more than does simply saying "I'm sorry" or attempting to made amends. The story of the German soldier and the Belarusian woman shows that forgiveness is a restoration that corrects the distortion in the relationship. Through forgiveness, both parties are equals again.

For the perpetrator, full disclosure is an act of disarmament, Müller-Fahrenholz writes—a willingness to finally confront the truth about oneself. It shines light on the unspeakable aspects of wrongdoing, which paves the way for atonement. The humanity of the other laid bare seems to spark responsibility in the listener and creates a catharsis and a way of moving forward.

James O'Dea believes that deep dialogue is the most powerful of healers because it dismantles both perpetrator and victim, allowing each to acknowledge the deep truth of an experience and their Bond in that history. O'Dea discovered the power of deep truthfulness to bridge the greatest of breaches during a social-healing meeting that included Mary Rothchild, the daughter of a Holocaust survivor, and Gottfried Leich, who had been a member of the Hitler Youth Movement in the Third Reich. Leich had been intensely fearful of revealing his history and facing his involvement in the Nazi movement, particularly in front of a Jew and the daughter of a Holocaust survivor.[22] He had dreaded this encounter more than any in his life.

Early in the dialogue Rothchild turned to Leich and said, "There are many people in my extended family who were killed in the Holocaust. What was your role in it?"

Leich admitted that he had beaten up people and set buildings alight on Kristallnacht, the Night of Broken Glass, in 1938, when Nazi storm troopers smashed thousands of Jewish homes, shops, and synagogues. "But I was only sixteen," he said.

Rothchild wasn't satisfied. "So, if you had been ten years older, would you have taken my relatives to the gas chamber?"

Leich took a very long time to answer and looked down at the floor. "I just don't know," he said finally.

Rothchild was utterly transformed by his answer. She had not expected such stark honesty from a person who represented the very archetype of otherness to her. With Leich's admission they both entered a completely new territory of truth: *Yes, I could have been a mass murderer.*

Leich became overwhelmed with self-loathing. He broke down and cried. "I am a grandfather, and my grandchildren are the grandchildren of the Nazis. I am in the abyss, the dungeon of history. I always will be connected with this."

His complete candor had released something inside of Rothchild, an emotional compensation she didn't know she had been seeking. She stood up, walked over to him, and took his hand. "Yea, though I walk through the valley of the shadow of death," she whispered to him, "I will fear no evil for thou are with me."[23]

Thompson and O'Dea both felt the space between Rothchild and Leich suffused with what felt like a divinely animated presence.[24] For Leich, the experience was a revelation. Surrendering his fear and facing his past truthfully had created a miracle; Rothchild, he said, had thrown a "bridge across the abyss" and invited him to meet her on it.[25] Through their shared history both felt the Bond.

According to Judith Thompson, *forgiveness* in Greek literally means "untying a knot," so both perpetrator and victim are free from hurt or shame, the legacy of the past, to carry on with their lives. Such is the power of deep disclosure or genuine confession that both sides are forever changed by the encounter. When the humanity of both parties shines through during forgiveness, a plan of restitution and new options for the future often simply present themselves.

Viewed from this perspective, a disagreement or wrongdoing is an interrupted connection, and forgiveness and restitution a reestablishment of that connection. As Müller-Fahrenholz puts it, transgression is a "sin against the whole" and deep truth an end to the "war of the world with itself."[26]

*　　*　　*

I run an ongoing Intention Experiment, testing the power of group thought in well-run, controlled scientific experiments and also informally in workshops and our Intention community. In our twenty-three studies to date we've achieved compelling results that demonstrate the power of thought to increase plant growth, purify water, and lower violence.[27]

The most interesting phenomenon to date is the effect on the participants in our larger studies. Our large online experiments have elicited many reports of ecstatic experiences, particularly an overwhelming and palpable sense of oneness. Although taking part on separate computers all over the world, the participants, who come onto our website at the same time, feel a sense of connection with other people in the experiment who are thousands of miles away. I have long suspected that there is a measurable long-term effect of this connection, just as there are long-term effects from meditation.

In September 2008 I ran an experiment with 15,000 participants from sixty countries, examining whether "group mind" has the power to lower violence and restore peace. The plan was to have readers all over the world join forces on our website to send peace to a particular war-torn area—in this instance, Sri Lanka.

In a survey I conducted of participants after completion of the experiment, some 46 percent said they noticed long-term changes in their relationships with others after the experiment. The group experience apparently helped them to feel more love in general, whether or not they knew the recipient. More than 25 percent felt more love for their loved ones or for people they normally dislike or argue with, 41 percent felt more love for with all those with whom they came into contact, and 19 percent found they were getting along better with perfect strangers

When asked with whom their relationships most improved, 38 percent said they noticed the biggest change in their relationships with strangers. The experience of connecting with thousands of strangers on our website and in Sri Lanka gave many people the ability to be more accepting of all strangers.

I have come to call this phenomenon the "power of eight." This type of bonding can occur in several minutes. During weekend workshops we divide the audience into small groups of eight and ask these complete strangers to send loving thoughts to each other. We've witnessed powerful stories of emotional or physical healing among both the senders and the receivers. Marsha, for instance, had developed an opacity in one cornea, largely blocking the vision of that eye. The following day, after her group's healing intention, she claimed that her sight in that eye had been almost fully restored. Many of the members of her group who had sent the intention were plagued with lifelong migraines or back problems, but they also reported feeling better.

Any possible healing effects may have to do with the powerful effect of community. During these workshops strangers begin resonating together as one; for instance, at a workshop in Holland I discovered that many of the groups reported having the exact same visualizations during their group intentions. In one group, which concentrated on sending intention to a woman with a bad back, she and a variety of members of the group all imagined the same inner vision: her spine being lifted out of her body and infused with light.

The feeling of oneness reported by the members of the workshops and the Intention Experiment community is an example of the resonance effect of pure connection. The simple act of belonging and spontaneously giving within a small group of strangers is so powerful and so satisfies our deepest longing that it heals both the healer and the recipient.

Recently, an American researcher from the University of California was conducting research on the Suya Indians of Mato Grosso, Brazil, in an attempt to determine how they use numbers. This group of Amazonian Indians are famous for their music; according to Anthony Seeger, a professor of ethnomusicology at the University of California, Los Angeles, their singing is used to create community, establish relationships and social identity, and also to formulate ideas about time and space.[28] Singing, to a Suya, is both hard and soft science.

Scientists who examine differences in number systems between cultures have concluded that many native cultures don't have language to describe quantities of things. For instance, the Piraha people use the same word, *hoi,* to mean "about one" and "about two"; the only difference is a subtle alteration in inflection. The much-studied Munduruku in the Amazon have words for numbers only up to five. This has led many scientists to examine whether human beings have innate numerical skills or whether number sense is a part of cultural conditioning.

On this occasion, the American researcher posed a numerical problem to a member of the Suya tribe: If you had ten fish and gave away three, how many would you have? The Suya answered without hesitation. As anybody in the village could tell you, the answer was thirteen.

In the Suya tradition, whenever you give something away to someone, the recipient pays you back double. So if he gave three fish to his brother, he said, his brother would have to give him back two times three fish, or six. Added to his ten original fish, this would give him sixteen fish. Once he deducted the three fish he gave his brother, he would have a net increase of three, or thirteen. So $10-3 = 7$ in Western mathematics transforms into $10 + (2\times3)-3 = 13$ in Suya mathematics.

The Suya was dismayed at the American version of the equation. "Why is it that 'giving' is always seen as a 'minus' for white people?" his fellow tribesman asked, entering the debate. "I know that you want me to use the minus sign instead of the plus sign, but I don't understand why."

This entire episode surprised Alex Bellos, the author of *Here's Looking at Euclid,* a study of cultural differences in mathematics.[29] He began his study with the belief that numbers are a universal language, a way we could, say, communicate with extraterrestrials, only to find that our basic understanding of arithmetical relationships depends upon cultural context.

This story reveals something very profound not simply about mathematics but also about how different cultures view relationships in general, particularly how we view ourselves in relation to other things. Our sense of mathematics very much depends upon how we define our world, and whether we view ourselves and all the things around us as individual entities separate from each other or inherently intertwined.

Many non-Western societies, including preliterate cultures such as the Australian Aborigines, the ancient Greeks and Egyptians, adherents of Eastern religions such as Buddhism, Zen, and Taoism, and a number of extant indigenous cultures, conceive of the universe as inseparable, connected by some universal energy or life force. This central belief breeds an extraordinarily different way of seeing and interacting with the world. We see the thing; they see the totality, the relationship between the things. To an indigenous native, giving is an act so inherently rewarding that it must always be denoted by addition, not subtraction. The most important aspect of a relationship is the plus sign—focusing on whatever is required to make the connection.

CHAPTER 11

THE NEW NEIGHBORHOOD

Henderson, the second largest city in Clark County, Nevada, has been burdened with heavy expectation ever since President John F. Kennedy, in a throwaway comment, referred to the sparsely populated upstart—a stone's throw from Las Vegas, newly seeded with casino money—as a "city of destiny." Within the next half century, as if following a presidential directive, Henderson swelled to the size of St. Paul, Minnesota, transforming itself into a middling American city, or as its website, Destination Henderson, puts it, "a diamond in the desert." "A place to call home" is the town's homespun motto, echoing the official welcome mat placed on the home page by the mayor himself.

Actually no one is particularly welcome unless you already happen to live within its gates, and even then, chances are that a massive wall stands between you and all your nearest neighbors. Henderson is home to Green Valley, the first of America's burgeoning number of "master-planned," gated communities, serving a population of 60,000—the size of many midsize towns—and constructed with the primacy of the individual specifically in mind. Walls of precise design and construction have been built between dwellings, at the end of backyards, between sections of the community, and, most of all, between the community and the outside world. Bans in place prohibit residents from altering the walls in any regard, even those on their property. Besides the gated entrance, high-end properties also come with their own security guard, and no one is admitted without a security check. Stores, parks, sidewalks, playgrounds, open spaces, even the local school all rest within its walled center, serving their exclusive community.

Green Valley is one of the world's fastest growing types of neighborhoods. Presently some eight million Americans live in gated communities; eight of every ten new urban building projects are gated, particularly in the West and South and in suburbs outside large urban sprawls. One half million of the country's gated-community residents live in California alone, where some 40 percent of new homes are built behind gates or some sort of security device. This trend is not unique to America. Gated communities are now popular in South Africa, where land developers first wall off an area, then fill it in with roads and houses; the Middle East, inside of which armored vehicles patrol to protect Westerners with oil interests; and the United Kingdom, where much of urban renewal in the Docklands, the new financial center of London, is taking place. Even populations in developing parts of the world, such as Mexico, Central and South America, and China, find the idea of walled towns and neighborhoods compelling. Nordelta in Argentina, the largest *barrio privado* in the country, even offers its residents their own exclusive hospital.

Although residents cite security as the main reason for living behind a gate, research into the effect of gated communities shows that they have a marginal effect on keeping crime at bay. The best two studies, carried out by the police in Fort Lauderdale, Florida, comparing the rates of all manner of crime before and after a neighborhood closed off its streets, found no significant difference in levels of crime against property or person. Auto theft, burglary, and other types of crimes at first drastically fell, but quickly returned to previous levels once criminals got used to getting around the gates.[1] The second study, which examined the crime rates of several closed neighborhoods with that of Fort Lauderdale as a whole, found that gates made no real difference in deterring certain kinds of crime. Although crimes against the person were lower within the gates, the incidence of burglary or car theft fell in the first year and then rose to equivalent levels of areas outside the gates.[2]

Recent incidents within Green Valley's walls include serial rape, domestic murder, various robberies, teen drug dealing and consumption, chlorine-gas pollution from a nearby industrial plant—in short, all the problems of ordinary, ungated suburban neighborhoods.[3] In fact even the most elaborate security in gated communities has not worked as well as simple Neighborhood Watch schemes, which have been shown to decrease robberies and burglaries by 24 and 33 percent, respectively, according to a study by researchers at Florida International University.[4]

Residents of gated communities point to crime as the ostensible reason for the walls, but the more exclusive high-end ungated areas of the country—where most gated communities are situated—already experience negligible crime. The real point of a gated community is to shelter its inhabitants from outsiders, write Edward J. Blakely and Mary Gail Snyder in their book, *Fortress America: Gated Communities in the United States*: "Traffic equals strangers, strangers are bad and bad means crime."[5]

The other point of a gated community is the keen desire for upward mobility and status; as Blakely and Snyder note, "They feed on aspirations for exclusion and the desire to differentiate." One builder told Blakely that what buyers want is a home that "makes a clear statement about themselves and their lifestyles." The most recently built communities, such as "The Enclave" in Green Valley, create especially elaborate entrances to emphasize exclusivity and status.

What many people claim to seek behind a locked gate is an old-fashioned neighborhood, that place where their kids can play safely in the streets, the parks and schools are safe, and the neighbors wave at each other over the garden fence. Yet that is exactly the reverse of what a wall achieves. A gated community is very like a state that has seceded from the Union, supplying its own services and security, answerable to very little outside its walls, encouraging its inhabitants to abdicate any civic responsibility to anything on the outside. It also reflects the relentless move toward atomization within our society, our present tendency to create smaller and smaller groups that are more and more homogeneous. What this means, essentially, is that gated communities and many modern subdivisions have transformed our concept of "neighborhood" to mean "exclusive country club." The purpose of our neighborhood now is to make the biggest "I" statement that we can.

In the main, our idea of community is now largely one that, like our relationships, must consist of sameness—a giant group of "I's"—in order to work. Social scientists like Robert Putnam distinguish between *bonding* social capital (interacting with people like us) and *bridging* social capital (interacting with people not like us). As he discovered in his study of the effect of ethnic diversity on trust and civic engagement in modern America, the more we are surrounded by people who are different from us, the less likely we are to interact with *anybody*—like or unlike us—and the more likely we are to "hunker down," as he puts it, with our immediate family in front of our own television set.[6] It's gotten to the point where, if we don't happen to live in a neighborhood of sameness, we don't want a community at all.

There is no evidence that sameness, in the form of a gated community, creates a better neighborhood or more "social capital," the sociological term for community spirit and togetherness; as Putnam points out, America possesses the lowest social capital in its history. The gate in fact prevents social capital from flourishing precisely because it encourages an in-group and an out-group.

The most powerful way to create a vibrant, open neighborhood is to move beyond the tendency to cluster together in similar groups and to find community in the space between, the space of interdependence where all of us join together in our common humanity and common purpose. One of the fastest ways to achieve this Bond is to form a community superorganism.

In 1954 twenty-two eleven-year-old Protestant boys from Oklahoma City, of similar middle-class backgrounds, unacquainted with each other, and carefully screened for psychological stability, boarded two buses bound for a 200-acre Boy Scouts of America summer camp near Robbers Cave State Park in Oklahoma. Observing all of this from his perspective as camp "janitor" was the architect of the study, Muzafer Sherif, a Turkish-born Harvard graduate, later credited with founding the discipline of social psychology. The boys did not know, in those days before informed consent, that they were about to become guinea pigs in one of the most fascinating and beloved psychological studies of group behavior of all time.

The boys were divided into two groups, and for the first several days the camp counselors (a team of psychologists, including Sherif's wife) encouraged them to engage in activities to solidify their bond with members of their own group. Each group was told to choose a name (they selected the "Rattlers" and the "Eagles"), create their group's own flag, designate places in the camp as theirs, write their own songs, and design practices and particular modes of behavior as exclusively theirs.

Each group was assigned separate living quarters far apart from the other, and at no point during this initial period was anyone allowed to meet members of the other group.

In the next phase of the study Sherif and his colleagues engineered situations designed to be highly competitive and frustrating in order to deliberately bring the two groups into conflict. They announced a general tournament of sports and other competitive games, with a trophy, medals, and eleven four-bladed Swiss Army–style knives as the much coveted prize for the winning team. Finally, after a day spent practicing, both groups finally laid eyes on the already loathed other.

Over the four days of the competition the staff manipulated the scorekeeping so that the scores remained neck and neck and ensured that the two teams were constantly made aware of the closeness of the race. Good sportsmanship gave way to name-calling, invectives, and refusal by every boy even to eat when a member of the other group was present in the same food hall.

After a while the experimenters didn't have to stoke the animosity between groups. Decking themselves out commando-style, the Rattlers raided the Eagles' cabin, turning over beds and ripping mosquito netting. The Eagles retaliated with interest; armed with sticks and bats, they threw all the Rattlers' belongings in a heap in the center of the cabin. Each team destroyed the other team's flag. On the day the Eagles won the tournament, the Rattlers raided and stole the prize knives. The growing animosity ended in a fierce fistfight that the counselors had to break up.

Having escalated such fierce prejudice between the groups, Sherif then experimented with activities to encourage the groups to commingle. But no amount of jolly, getting-to-know-you evenings, movie nights, or festivities on the Fourth of July seemed to lessen the tension.

Sherif then created a series of crises in the camp that could not be solved without the resources and participation of all the boys in both groups. After the drinking water suddenly dried up and the boys discovered a large sack stuffed into an outlet faucet of the water reservoir, both groups had to work together to figure out how to clear it. The Rattlers and Eagles were also enlisted to pull a rope tied to clear a partly cut-through tree that may have posed a danger, and then to help with a truck carrying food for both groups that had gotten stuck in a rut.

Once the water began flowing again, the Rattlers allowed the Eagles to drink first from the fountain, because the Eagles hadn't brought their canteens and were thirstier. After the groups worked together to finance a movie, the boys began eating together in the mess hall, Rattlers freely mixing with Eagles. On the final day of camp the boys unanimously voted to travel together on the same bus. Rattlers and Eagles sat together, arms draped around each other. At a stop on the way the leader of the Rattlers spent the five dollars he'd won on a bean-toss contest on malted milks for all twenty-two boys.[7]

William Golding described his vision of the inherent "darkness of man's heart" in *Lord of the Flies*: when the veneer of civilization is removed, even children are capable of turning savage. Sherif's study demonstrates the opposite. Schoolboys indeed are fully capable of being cruel and bullying each other if they are placed in opposing groups and forced to compete over scarce resources. But when the Robbers Cave children were given a common goal and purpose larger than themselves and their group they readily put aside their differences to work together cooperatively as a superorganism.

The Sherif study, which was repeated several times, is considered a classic test of the *contact hypothesis,* which was devised by Gordon Allport that same year. A towering figure in psychology and one of the founders of personality psychology, Allport believed that managed contact between members of different groups is the best means of reducing prejudice. His theories influenced the Supreme Court in its landmark decision ending school desegregation, *Brown v. Board of Education of Topeka.* Allport's theories also spawned some spectacular failures to end racism in America, such as the busing of American black schoolchildren into white schools during the 1960s.

Robert Putnam's evidence would appear to counter Allport's conclusion that contact between groups leads to solidarity and trust; Putnam found that the more ethnic groups Americans are surrounded with, the higher their prejudice and distrust. However, Allport specified four specific conditions for ensuring that contact between different groups works: equal group status in the situation; intergroup cooperation; the support of authority figures; and finally and most important, a common, overarching goal.[8]

Psychologists call this a superordinate goal, a goal achieved only by strong cooperative teamwork. Engaging in sharing and teamwork tends to transcend differences because it emphasizes the very heart of humanity: *we are all in this together*. And if we are all in this together we are no longer competing for scarce resources.

Although criticized in some quarters as resting on the underlying assumption that all people are fundamentally similar,[9] the contact hypothesis has been tested in a variety of settings and found to be a powerful tool to lessen discrimination against many groups, from Protestants and Catholics in Northern Ireland to homosexuals at universities. Sports teams, management teams, schools with new members, and even prisons make use of superordinate goals to alleviate rivalry and encourage team spirit. In 2006 a review of 525 studies of the contact hypothesis in action confirmed that contact between different groups of any variety successfully reduces intergroup prejudice and increases cooperation, particularly when Allport's four conditions are met.[10]

Don Beck was a graduate student of Muzafer Sherif, and he used the lessons of Robbers Cave to create superordinate goals as a means of ending political conflict. It was Beck who first came up with the idea of using South Africa's entry into the World Cup rugby playoffs in 1995, depicted in the film *Invictus,* as a means of creating nation-building euphoria to unify a country emerging from apartheid. Beck had a special fascination with the psychology of premier-league games, and through his experiences working with the Dallas Cowboys and the New Orleans Saints he had developed a belief in the power of sports as a peacemaker. This was a bold idea, given that the Springboks, the South African rugby team, were the very symbol of apartheid. Rugby was considered a white man's sport. Virtually all players were Afrikaners, the white pro-apartheid minority; rugby coaches even shouted out plays in Afrikaans. English-speaking and black players seldom made the team, and consequently the black population in South Africa actively boycotted the sport.

In 1995 Beck presented Kitch Christie, the Springboks coach, with a paper titled "Six Games to Glory," which detailed a series of psychological strategies that would help transform the team from underdog to world-class contender in the games leading up to the World Cup. Besides the strategies for winning the game, Beck's paper included ways that the Springboks could become a focal point of pride for the fledgling country and connect the township blacks with the Afrikaners.

Beck's paper offers many strategies that can be used to create superordinate goals in other areas. He suggested that the Springboks adopt a *collaborative* or *common identity:* the green and gold colors of the team shirts and a team song with a Zulu drum to lead the team and arouse the crowd. He advised Christie to have the team sit together and watch films such as *Hoosiers* and *Chariots of Fire* to help establish a "mystical brotherhood," the sense that the team *stands together as one family,* with a bond greater than their loyalty to themselves and *a cause to fight for.* Beck arranged for the team to visit Mandela's tiny prison cell at Robben Island to emphasize *their larger role in their country's destiny.* Above all, his exercises were meant to develop a sense that each member of the team faced *a life-defining moment* requiring that they *pull together as one.*

As the games progressed, Beck's superordinate goal began to infect the country; young blacks from the township tore down antirugby signs and hung photos of their Springbok heroes. During the World Cup, which the Springboks went on to win, Mandela was persuaded to appear in a Springbok green and gold shirt, which had always symbolized his oppressors, as a tangible sign of unity and forgiveness.

To Beck, creating a superordinate goal is one of the best ways to achieve peace in areas of political conflict. In his work he often meets with both sides in an area of discord and shows them a positive vision of future possibility, but one that requires that both sides work together and use their common geography and resources to create a solution for all who live there.

Beck likes to talk in sports metaphors. "You focus on relentless pragmatism," he says in a heavy Texas drawl, "the progress that can be made to move the game forward." Recently he presented the Arabs and Israelis with a plan to make occupied Palestine "the Hong Kong of the Middle East," an affluent society with both sides sharing resources for services such as education and health care. Presently he is consulting with both sides to work out the details and timeline of how to achieve this kind of society within thirty years.[11]

Creating a common identity and working together for a superordinate goal was also crucial to the survival of the Chilean miners during the seventy days they were trapped below the Atacama Desert after the Copiapo mine collapsed in August 2010. The shift supervisor and de facto leader, Luis Urzúa, used a variety of tactics to create a collective identity: strict sharing of resources; a one-man, one-vote democratic decision-making process; a unifying name for the group (*Los Trenta y tres*, The Thirty-three)—all of which helped to create a sense of *one for all, pulling together against impossible odds*. At the same time Urzúa constantly reinforced the fact that survival wasn't just personal or for the good of the group. He hung up the Chilean flag and frequently led the men in the Chilean national anthem. He created in his men the sense of their place in history: their *survival was necessary for the good of their country.*[12]

From a scientific point of view, the true power of leaving our small space of individuality and coming together as a group to achieve a superordinate goal stems from a collective resonance effect. Just as brain-wave entrainment can occur between two individuals, so it also gets established between group members working together. The electrical activity of each individual in the group begins to resonate on a common wavelength—a choir perfectly in tune. Like a group of electrons that begins to vibrate as one giant electron, the group creates a resonance that magnifies the individual effect.

Psychologists at the Max Planck Institute for Human Development in Berlin and the University of Salzburg wished to examine whether our brains act "in tandem" with others when we're engaged in a common purpose. Although some research had been done with fMRI, no one before had examined simultaneous brain-wave activity between people carrying out the same task together. The German scientists were inspired by recent studies examining the brain-wave rhythms between two people when they socially interact, demonstrating that one type of brain wave is associated with independent behavior, while another is shared by both parties when the behavior is coordinated.

The German and Austrian scientists decided to study the brain activity of pairs of guitarists playing a short melody together to see to what extent cortical activity is synchronized between people when they are, as they put it, "swinging in concert." The scientists placed an EEG cap on each of the musicians and recorded their brain activity.

Using special algorithms to analyze the brain activity of each person individually and in relation to his partner, the scientists found that the brain waves of each pair were highly synchronized and "in phase"; that is, the waves peaked and troughed at certain key moments, when they were practicing setting the tempo with a metronome and then when they began their coordinated play.

In fact entire areas of the brain had synchronized patterns, with the frontal and central regions the strongest, but the temporal and parietal regions also showing high synchronization in at least half the guitarist pairs.[13] Because the parietal regions govern our sense of self in space, the musicians' synchrony suggests a move toward oneness. We move beyond the self when we are working to create something together.

This study has vast implications, considering that so much of our interaction with the world consists of synchronized and goal-directed actions with other people. The researchers concluded that whenever people do things together in a synchronized fashion, their brain waves must follow suit. Brain-wave synchronization may even help to maintain interpersonal relations between people, given that they play such a vital role in early social development. Like a jazz group working together as a superorganism to produce a common sound, we get on each other's wavelength whenever we're working together to produce a common result. Ultimately this is likely to be the basis of all successful group relationships. We are able to get along with each other—no matter how different we are—simply by sharing an activity or goal.

We also experience a powerful physical boost by being part of a superorganism. One of the most compelling studies of this phenomenon involved one of the greatest of British university traditions: the rowing team at Oxford University, known to be fiercely competitive, particularly against their chief rivals at Cambridge. Anthropologists from Oxford's Institute of Cognitive and Evolutionary Anthropology asked a group of the Oxford rowers to work out in a "virtual boat" housed in a gym and used for normal training. In each test they were to row continuously during forty-five-minute sessions, first as members of a team, and then as individuals.

After each session the scientists measured the rowers' pain thresholds by measuring how long they could tolerate an inflated blood pressure cuff on their arm. Exercise has long been known to increase a person's ability to tolerate pain. Although the rowers evidenced increased pain tolerance after every session, they had significantly greater tolerance to pain after the group training compared to exercising individually.

The scientists concluded that although all physical activity results in a release of endorphins, one of the body's feel-good chemicals, the synchrony of the shared physical activity appeared to create a ramped-up endorphin release, which may have something to do with communal bonding. Emma Cohen, the lead author in the study, noted that "synchronized, coordinated physical activity may be responsible" for the phenomenon. The rowers created a "field" that magnified individual efforts and overrode individual limitations.[14] Within the field the whole was greater than the sum of its parts.

When we do things in groups, the rush of *we're all in this together* elation that we feel actually allows us to resist difficulties, including pain. This offers proof of the old adage that there is power in numbers and also explains why we feel something extraordinarily akin to magic in groups working for a common purpose. We move outside our own individuality and into the space of the Bond.

Scientists now understand that neurons become more efficient and operate as a unit when they are repeatedly and persistently stimulated together: *neurons that fire together wire together*. It may also be true that people who fire together wire together. When we work with others for a common purpose, we very quickly and literally get on their wavelength. All of this suggests that coming together in small groups with a superordinate goal provides a social cohesion beyond money, job, or size of property. We may be at our happiest when neighbors are helping neighbors—as neighbors used to do, for instance, when raising a barn.

There are many names for activities like barn raising; the Native American Cherokees call it *gadugi*, the Finnish *talkoot*, and some Americans call it a *bee*. The idea is to pitch in together for a common cause—whether making a quilt, husking corn, or raising a barn—that is difficult or tedious when carried out on your own. In Norway people engage in *dugnad* to garden in common greens or to help build houses, and some organizations arrange an annual *dugnad*. Other communities work together to create something of value for the entire community, as they did not long ago in the tiny town of Tailholt, Oklahoma, one of the fourteen counties of the Native American Cherokee Nation.

Tailholt is one of the forgotten neighborhoods of America. The town sits in a remote corner of southeastern Oklahoma, a stone's throw from towns with names like Bunch and Greasy and Tenkiller. Tailholt's own name, derived from the pioneer practice of traversing a flooded river by persuading a horse to ford the rushing water while you hold on to its tail, suggests a place hanging on for its very survival. One-third of its population of 42,000 consists of Native American families, with an average per capita income of $27,000 and a house worth about $60,000. Those businesses that exist in Tailholt mostly cap employment at minimum wage. With a cemetery within each mile of a twelve-mile radius, the most prosperous activity in Tailholt is laying its inhabitants to rest.

The citizens of Tailholt had been trying and failing to get fresh water every year since 1999. Many members of the community had constant problems with their water sources. Wells ran dry, taps had low pressure, and water was contaminated or just smelled or tasted bad. The US's Environmental Protection Agency has strict guidelines about coliform bacteria, which give an indication of the amount of harmful microorganisms in the water, and the maximum number of coliform bacteria per milliliter that constitutes water fit to drink. Certain forms—called fecal coliform bacteria—can make people ill. Fifty-eight percent of Tailholt households failed the coliform test. Every annual application to the national Indian Health Services to fund a new pipeline through a grant available to Indian communities had been rejected on the grounds of expense. There was just too little federal money to go around.

Tailholt residents were also desperate for a large community center as a central meeting place for organized activities. Again it looked as though the high cost of the project would never be picked up by the federal government.

In 2004 the Cherokee Nation formed the Cherokee National Community Work Projects to provide small amounts of funding to help Native American communities like Tailholt when federal funding isn't available. The Cherokee Nation's Community Organization Training and Technical Assistance (COTTA) scheme was also set up, to teach communities how to band together and maximize any little money they could get.

When the Tailholt community failed to secure the federal funds for their projects, they met with Billy Hix, director of the Cherokee

Nation's Engineering and Sanitation Facilities Construction Program within COTTA, who convinced them to be active participants in the construction of the waterline. The community needed points to help them qualify, and one measure of adequate points was how much of the project they were willing to do themselves.

The Tailholt community began holding regular meetings, with up to two hundred people attending. Of those a core group of thirty agreed to work on the community's goals of a town meeting center and a fresh water pipeline. They agreed to provide most of the manpower to dig and bury the ten-mile pipeline into a four-foot ditch—a process that would take four to six months—with the county's water department overseeing the project and providing technical assistance.

Although the water line would have cost some $579,000 to lay down, more than half of that cost was deferred because the citizens offered to do the digging themselves and provide their own equipment. By subtracting what the labor would have cost, Hix was able to drastically reduce the federal funding necessary to get the project approved.

Because of Tailholt's willingness to invest in sweat equity to build the community center as well, the Cherokee Nation provided $72,000 to pay for the basic materials for the 3,700-square-foot building. Tailholt's request for small amounts of federal funding again made it through the fierce approval process because volunteers offered to carry out the construction work for free.

The community began scouting out potential locations for the building, but once again they were faced with money problems. Where would they get the funds to buy the land? At one of the evening meetings eighty-year-old Pauline Sanders stepped forward. She had the perfect site for the building on her substantial acreage, and she was willing to donate nearly five acres, with the proviso that the center offer a literacy program for children and a nutrition program for the elderly.

The building and pipeline were both operational by 2006. Tailholt had clean water and a community center with a library, with free computer use, and a place for everyone to meet. But the bigger payoff was the effect on the community of engaging together in a common goal. Before the building work began, residents had felt isolated from each other, but during construction a large number of the men showed up to the building site with their hammers, squares, and levels, while Pauline

Sanders and a number of the town's women gathered at Pauline's house to cook lunch. "The whole process of getting this community building started has brought our community together," said Jeremy Marshall, president of the Tailholt Community Organization.[15]

As soon as it opened, the center became a fulcrum for the entire town. Further plans are afoot for a playground for the children, Cherokee-language classes, an afterschool program, and other activities for young and old. "It puts the future for the kids in the community," says Sanders. Community involvement became infectious in Tailholt; residents began to volunteer for the rural fire department and fundraising.

Many of the citizens believe that the self-help nature of the project was the key to its success. "It motivated the people and it makes them appreciate working for something instead of just having their hand stuck out for a handout," says Lynette Studie, the chair of Tailholt's fundraising organization. "It's involved more people. It's the only way to do things."[16]

The lessons from Robbers Cave and Tailholt are not simply a blueprint for reducing prejudice or getting a community center built. They also offer a simple means of creating a New Neighborhood, tearing down the walls between people as they did in my own community.

There were only two times when my neighborhood in suburban London came together for something besides a perfunctory gathering. In both instances it happened out of necessity: we were under siege. With no warning, Orange, a British cell phone company, announced its intention to install eight cell towers in our community, with one planned right on our block, directly across from our younger daughter's bedroom window. The majority of members of our town, in particular our immediate neighborhood, were alarmed about the potential detrimental effects of a cell tower on our health, especially that of our children, as well as on our property values and the general aesthetics of the neighborhood.

Within days we were party to the most extraordinary social transformation. A group of us met at our house one evening. Over tea and cookies we put together a comprehensive plan to form a "housewives'" brigade to protest Orange's plans. Faced with a common purpose and little time, we needed no one in charge. Instinctively we knew to play to our resources and strengths. Everyone simply volunteered to take on whatever was necessary, and the division of labor occurred automatically.

One of the businessmen took it upon himself to study the law, to see what grounds we could use to protest. Because the health issues believed to be linked to cell phone towers are controversial and were not at the time considered valid grounds for a reasonable objection, our greatest challenge was defining what exactly we were objecting to. We had to rely on other factors—aesthetic issues, or threats to the safety of pedestrians in wheelchairs or mothers with strollers—to make our case. Several neighbors scouted around and eventually located sparsely populated sites in the area where phone towers could be placed as a reasonable alternative.

A neighbor approached the headmistress of the Catholic school on our street and all the ministers of all the other local churches for their support. Our next-door neighbor built and painted a giant luminous orange box in the dimensions of the proposed tower, and parked it on the proposed location to give the neighborhood a visceral idea of the sheer dimension of this proposed tower and exactly how much of an unsightly and cumbersome impediment it would present on our side-walks. As owners of a small publishing company, my husband and I volunteered to design and produce posters, fact sheets, and sample let-ters to our local council, and petitions to Parliament.

We parceled up the area and took turns distributing leaflets. Some of the women stood outside school gates and knocked on the doors of every unit in every apartment building; others contacted our local member of Parliament. One of the families with a distant link with Orange organized a meeting with a company manager and invited our parliamentary representative, during which we discussed our objections and proposed a reasonable alternative. We were clear that we meant business if they refused to take our concerns seriously.

Within a few weeks Orange withdrew its petition.

Several years later they were back. This time, most cynically, they made their bid over the summer months, when most people were on vacation. We were alerted to their renewed campaign only after my husband noticed a small poster that had fallen off a tree across the street. Nonetheless within a few days we resurrected our local email list, updated and reprinted the petitions and the fact sheets, and this time enlisted a few teenagers on the block to pass out the material. Within a month, after hundreds of letters of protest had been sent to our local council, they again turned Orange down.

Although Orange appealed, ultimately the company decided that fighting this well-organized resistance wasn't worth it. Eventually they withdrew their appeal and joined up with another cell-phone provider for power in our area. At this writing it appears that what was largely a ten-person-strong housewives' brigade has chased away one of the giants of British industry permanently.

Although it was gratifying to see the power of a tiny group of committed citizens in action, what most interested me about this episode was the effect of this crisis on our relationships. We have a mixed community; across from private houses built in the 1930s stand what were formerly public housing projects, a relic of 1960s social engineering, when the town planners knocked down many houses and replaced them with clusters of apartment buildings for working-class people. Since the reign of Margaret Thatcher these council apartments have been privatized and the inhabitants allowed to purchase their own flats. Consequently the people on our street are all homeowners, though some homes are far larger than others. The shops within the common grounds are run by hardworking, ambitious South African and Indian immigrants; our little grocery story is owned by an Indian widow who works all hours to put her children through private school. Nevertheless, to many of their more affluent neighbors, the apartments and the shops near them were no-go areas—Beirut, as one of my neighbors once put it in half jest.

During our crisis these same neighbors were forced to interact with Beirut, put aside their differences, and relate to each other on a deeper level. In this emergency we found the soul of a community we never thought we had.

Individuals within a community can even make their financial savings a superordinate goal, as it is for many people in Okinawa. Yoko Masashiro, a fifty-eight-year-old widow, has owned a health food store in her hometown of Naha, the largest city in Okinawa, since her husband died in 2000. The name of the shop is Ichariba, which is Japanese for a place where people meet and become family. "Ichariba" could also stand for the way that Yoko handles her finances. Every month a trusted friend comes to her house, and she hands him ¥50,000 (about $500), in the full expectation that at some point within the next year that amount will be handed back to her tenfold.[17]

Despite the American influence on this tiny island, with thousands of American troops stationed there since World War II, Okinawa still maintains its ancient customs, including a unique system of mutual aid, or *yuimaru*. Their informal system is called *moai*, which is a little like a savings-and-loan club organized by friends who attend weekly or monthly meetings for a set period of time. The currency is not only friendship, but also, in the case of a *tasukeai moai*, a pool of money to which everyone contributes for a set period of months, depending upon how many are in the *moai*; for a group of ten people, the *moai* will carry on for ten months.

In Yoko's *moai*, each month everyone puts in ¥50,000, and the money is portioned out to one of its members in turn, with the understanding that at whatever stage you take the pot you will begin to pay interest of ¥2,000 for each of the remaining months of the arrangement. The person who opts to receive the money in the first month will receive ¥500,000 and then pay the ¥2,000 for the next nine months. The second person who opts to take the full pot in month two will receive ¥500,000, plus the ¥2,000 from the first person's interest, and in the third month, the third person will receive ¥500,000 in the pot plus the ¥4,000 in interest from the first and second recipients, and so forth. The first person to benefit from the money will pay the most interest, and the person who patiently waits to take the final pot will receive the most money (¥518,000); if two elect to receive money in the same month, the recipient will be determined by the toss of a coin. Sometimes the sum is nominal, with no interest charged; the *moai* is just an excuse for a monthly get-together with friends, as it was with Yoko as a young married woman. She set up an ongoing *moai* with her college friends simply to share an evening away from their busy family lives.

Moai were born of the Okinawan deep distrust of banks and their complex procedures, and to this day many Japanese in rural areas would rather fund their cars or houses through their friends and the *moai* pool than a bank loan. Yoko believes that a *moai* is a better financial system than a bank, particularly when the interest rate is low, but also because it is so mutually supportive: you help others when they need money, and they do the same for you.

A *moai* is a bit like a Public Goods game, which depends entirely on trust and everyone's willingness to continue to pay fairly every month. When deciding to join a *moai* members look to the trustworthiness of the organizer, who will be held responsible if something goes wrong. In fact on mainland Japan, where the practice is less common, *moai* goes by another name, *tanomoshi-ko*, which roughly translates into "reliable group." It goes without saying in Japan that each member of a *tanomoshi-ko*, particularly the organizer himself, will be a strong reciprocator.

Another way to forge a strong community is to create an all-in-this-together sense of collective pride. Stewart Wolf's study of Roseto, Pennsylvania, reveals that the town had become a highly cohesive cultural community because of certain specific practices among the townspeople. Besides their civic pride in the beautiful town itself, residents shared a certainty about the future; young men knew they would work in the mines and the women understood they would work in the local blouse factory. *You knew where you stood.* Most families had three generations living in the same house, and virtually no one was on welfare; the townspeople ensured that no one was left out. Another aspect of the collective caring that especially stands out was the fact that people did not strive for distinction at all costs, particularly at a cost to their neighbors. Such was the sense of solidarity that ostentation was strictly discouraged and jealousy consequently minimized. Although rich and poor lived together, side by side, the rich did not flaunt it. Roseto was flushed with a clear sense of common purpose.[18]

A similar situation used to occur in Syria. Nour Hakki, a thirty-year-old translator in Damascus, remembers hearing stories about her neighborhood in the days of her grandmother. Every house, built of stone or brick, was deliberately kept simple outside. Inside a corridor gave way to a lavish garden within, with trees and fountains and flowers surrounding a central courtyard. Gold jewelry was worn only inside as well. The idea was to avoid outer ostentation so that you would not "break your neighbors' hearts." If someone didn't have the money to buy a house, the neighborhood chipped in. If you were cooking something and your neighbor smelled it, you felt compelled to offer him some. Hakki mourns the passing of these old ways. Islamic countries are becoming Westernized; women want to wear only Chanel and Gucci now. "Ostentation," she says, "is a sign that God is blessing you."[19]

The social psychologist Willem Doise observed the very human tendency to place ourselves into subgroups and suggests that one way we can come together is by "cross-cutting categories," attaching ourselves to more than one group. That practice not only reduces the prejudice against out-groups but also tends to stop people from making comparisons. It reduces our need to focus on a single factor—religion, sexual identity, politics, or even socioeconomic background—in order to feel that we belong. People who belong to many groups create a "superordinate" identity for themselves, which in itself has been shown to reduce prejudice and fear.[20]

In fact the healthiest situation is not strong association with a single group, but broad diversity. Jolanda Jetten, a social psychologist at the University of Exeter, has examined many social networks and discovered that the most well-adjusted first-year university students and those with the least amount of depression are those who belong to the most groups.[21] We are happiest when we define ourselves according to the largest umbrella, just as the townspeople of Roseto did not cluster together because of income or religion or political affiliation. All were happy to proudly stand under a single banner—as Rosetans.

Discrimination doesn't require conflict or indeed much besides the flimsiest designation of otherness. As the American psychologist Henri Tajfel demonstrated, when a batch of adolescent boys were told that certain others had scored the same score as they had on a computer task, they began to band together and discriminated against those who hadn't achieved the same score. Difference of any sort that gets emphasized is enough to create a "minimal group" and, consequently, an outgroup.[22] All it takes is a wall, no matter how insubstantial.

Perhaps the most important way to reestablish the Bond within our neighborhoods and even our societies is to broaden the very definition of who we are. In his study of religious diversity, titled *American Grace,*

Robert Putnam discovered that America is becoming increasingly tolerant of religious diversity and far more disposed to having family members marry outside their own church. In the case of religion, contact and familiarity between dissimilar groups did in fact breed acceptance, even if ethnic diversity does not, at the moment.[23] That suggests that our ethnic identity is more firmly fixed and exclusionary than our spiritual identity, but also that acceptance and cooperation can be cultivated—and restored.

Once you remove the competitive nature of human groupings, people begin to flourish. Like overlapping molecules, we can learn to connect again and reclaim our natural way of being by creating a larger, all-embracing identity, a bigger definition of who "we" are. The more groups you can label as part of yourself, the more people you will embrace.

CHAPTER 12

PAY IT FORWARD

It was 1998 and a good year to be a student of computer science at the Berkeley campus of the University of California. Companies in Silicon Valley were in the middle of the dot-com boom and so short of manpower to feed the incessant hunger for web marketing that recruitment officers had resorted to scouring campuses and hiring undergraduates. America's five top accounting firms were there, attempting to lure new blood from Berkeley's School of Engineering and Computer Science and the Haas School of Business. As a computer science major with a minor in philosophy, Nipun Mehta had been an obvious early target

Mehta was a high-octane prodigy who had begun computer programming at the age of fourteen, entered Berkeley at sixteen, and by his junior year had already been snatched up for a part-time position at Sun Microsystems, which picked up the tab for his college tuition. Within a few years of graduation, he was being paid a six-figure salary, with stock options and sign-on bonuses thrown in. Like many of the other successful young turks in the Bay Area, he was riding the wave of the vast potential of the Internet. Almost every one of his programming friends from Berkeley was growing wealthy overnight, securing big stakes in initial public offerings in businesses that essentially sold nothing but promises. Some of them quit their jobs as programmers to become traders on the wildly speculative market. The entire topic of his friends' conversations seemed to concern how they were spending this enormous sudden windfall—who had bought the next house, the latest BMW, whose options in IPOs were better than others.

In a sense Mehta himself had been programmed for the boom as a self-starter with an obsession to succeed. Despite his age, as the youngest of his year he had graduated high school at the top of his class and as the star of local tennis tournaments. When those tournaments began to feel limiting, he enrolled as a full-time student at a community college during his final year of high school in order to play in a more competitive circuit. Besides being a wizard with technology, Mehta was driven by the prospect of financial success, and he brought his passion for winning to the stock market. While still at Berkeley he began trading stocks on the Internet, playing tens of thousands of dollars every day, refusing to quit until he'd turned a loss into a profit.

In the middle of 1999 Mehta began to have an uneasy sense about the get-rich-quick culture around him, particularly its effect on himself. At times he felt as though he were drowning in a sea of greed. Occasionally, he was reminded of an experience that had profoundly affected him as a teenager. By chance, on a long flight from Japan to India, he and his brother were seated next to a Japanese man named Shin. What began as a pleasant greeting became a compelling conversation that lasted the whole of the ten-hour flight. Shin was a physicist with a personality a lot like Mehta's, a workaholic whose twenty-hour workdays were fueled by fifteen cups of coffee. The year before the plane trip he had been diagnosed with late-stage prostate cancer and given only two months to live. Shin realized that his life choices had contributed to his situation, primarily what he described as his Japanese-style machismo: the need for relentless pressure and absurdly long hours at work. He had embarked on an intense course of inner and outer changes that ultimately brought about a spontaneous healing. More important, it changed his fundamental view about what exactly he was here to do. As Mehta began working his own eighteen-hour days, he kept reflecting on Shin's words. He did not want to turn sixty only to realize that all he had to show for his life was a large bank account.

Another profound lesson was waiting for him in India when visiting an old school chum. He'd taken a ride with his friend on his motorized bike, and as they belted along over the uneven pavement and potholes of Mumbai's back streets, Mehta was overcome by such uncontrollable nausea that he asked his friend to stop so he could vomit on the edge of the crowded street. Suddenly, a dirty, emaciated old street vendor appeared, riding past on his bicycle. When he saw Mehta, he stopped, slowly reached into a sack, produced a lemon, cut it in half, and handed half to Mehta, gesturing for him to suck on it in order to stop the nausea. From the look of him, it might have been his last lemon, yet he was happy to share it. Without speaking another word, the old man got back on his bicycle and rode off. He had showed up out of nowhere in time of need, carried out an act of kindness and then just disappeared, without requiring any credit. Mehta reflected deeply over the incident. If the roles had been reversed, he wasn't sure that he would have done the same. And how different it was from the philanthropy that went on back home in Silicon Valley, where givers were as ostentatious about their giving as about every other aspect of their lives.

Mehta returned to Silicon Valley determined to rethink his own life. He approached his friends at work to create a "donation club," to pool a small portion of their large incomes each month for charity or to gather together on Saturdays to make a giant stack of sandwiches and share them with the homeless.

One day in early 1999, Mehta and his friends showed up at ACT For Homeless in San Jose to offer help. When they got there, this group of young Silicon Valley engineers took one look at the shelter and had another thought. What the shelter really needed more than another set of helping hands was a website. At the time such expertise cost $10,000 or more—well beyond the shelter's budget. If they had a website, Mehta told the shelter's head, ACT could let people know of its presence and its services. A website would quickly attract donations from all around the Bay Area, far outstripping whatever Mehta and his friends were prepared to give. The group quickly knocked together a website for the shelter, which, as they had guessed, immediately enhanced its public presence. It also sparked a big idea.

One evening Mehta invited twenty of his young professional friends to a gathering at his home. He wanted to do nothing less than to shift the Silicon Valley culture of greed and materialism, he told them, though that was an impossible task. What they could manage, though, was to change that culture within themselves. Mehta invited them to carry out an experiment in the joy of giving on their patch of the West Coast, a part of California where it appeared that some of the most selfish people in America lived. Silicon Valley had the highest per capita ownership of Ferraris and the lowest percentage of charitable donations in the country.

Mehta had in mind starting a charity that would gift the skills of its volunteers to empower other charities and nonprofits with websites. With a few deft bits of programming they could transform a nonprofit into a tech-savvy organization, which could then effectively reach out to the community.

Mehta was clear about his purpose. "We're not here to entertain ourselves," he said. "We're not here to network." Ostensibly he wasn't even doing this for the charities themselves. The idea was to ultimately create a contagion of giving, by first creating the contagion within themselves. Trishna Shah, a friend of Mehta's brother in the midst of her first summer at Berkeley's Haas Business School, remembers that first meeting clearly. "It was all about giving, how the act of giving would transform us," says Shah. "We wanted to create an opportunity to transform internally." In Mehta's view of personal transformation, colored so much by the Internet, their efforts would quickly go viral, creating a slow but steady seepage of giving into the community.

Mehta formally incorporated CharityFocus in 1999. Its initial plan was to provide services for free for any nonprofit community service organization in need of help with internet technologies. CharityFocus was to be run entirely by volunteers, with no full-time staff, so that every member would donate his time in an unconditional way. Without salaries, all would be equally valued, without a hierarchy of titles or assumptions about which contribution was the most significant. They would accept only spontaneous donations and never seek to fundraise. If they were short of funds, they would work in creative ways with whatever they had. Their true aim was ultimately to focus on the small—on changing themselves, not the world.

One of their first ports of call was Toothmobile, owned by Mike Reza, whose small, cramped offices in San Jose offered dental care for the poor. Reza was passionate about his work. He drove a trailer to different sites, including poor schools, offering low-cost dental care. The biggest problem Reza faced was not volunteers but customers. No one knew of Toothmobile's existence. He needed to raise awareness but didn't know how, and a website was completely out of his reach. Mehta and a few CharityFocus members who had worked on Netscape built Reza's website for free, and then observed as the website transformed Reza's charity. Their small contribution, a matter of a few hours customizing a web template, created a completely different landscape for Reza's project. CharityFocus went on to build more than 5,600 websites for all manner of charities, from soup kitchens to international aid groups like Airline Ambassadors, which recruits frequent flyers with a large number of air miles to hand-deliver aid to children, provides free flights for orphans to travel to their new families, and flies ill children to get medical aid in developed countries.

In 2001 Mehta quit his job at Sun to become a full-time volunteer and turned to the Internet to spread his message. He and his team wrote the Daily Good blog, posting positive stories and quotes on websites, and created a website called helpothers.org, a kindness portal that "tags" people with acts of kindness and Smile Cards. After doing a selfless act for someone, you leave behind a card asking him to pay it forward, repaying this kindness with another act of kindness. He created KarmaTube, a YouTube of "do something" videos suggesting small actions that individual viewers can take, highlighting the work of "change agents" and micro-communities of volunteers. Their Wednesday programs created a home base in individual communities for people to meet for local meditation, discussion and dinner.

Mehta has designed Wednesdays as an experiment to test the inherent trust that develops between people who are generous with each other. Every Wednesday, Mehta's parents open the family home in Santa Clara to strangers. At this writing they have fed more than 25,000, with forty to sixty new people arriving each week, and similar gatherings now occurring in twenty-four locations around the world. The attendees come from all walks of life; millionaires sit next to the unemployed. Nevertheless, all are given a chance at the microphone, after a reading starts the discussion each week. Generosity is a great leveler. Everyone understands that in this atmosphere they are valued and seen as one.

"When generosity is the basic social capital, you see things from a broader perspective," says Mehta. "You come from a different place of openness. You're more likely to see multiple views. It deepens trust. The cup of gratitude overflows, and turns into action in so many ways."

In thirteen years of weekly "Wednesdays" in Santa Clara, there have been no major difficulties, disagreements, or theft. The only argument concerns who will do the dishes. There is no donation box, so people find creative ways to give back. One Wednesday attendee left a shoe rack with space for 100 pairs so that the other visitors, who always enter the house in stockinged feet, would have a place to leave their shoes. Mehta keeps tweaking the experiment. Most recently, he created an "open source" library of his parents' living-room bookshelf after filling it with inspiring books and announcing that people can take or leave books whenever they want. The system is operated completely on trust, with no sign-out system. Thousands of books have traded hands, and the shelves are now crammed with many more books than Mehta started with.

Mehta throws out fresh challenges to each new Wednesday group. *Reach out to the neighbor who has irritated or ignored you. Write them a card, take dessert over, haul their garbage to the dump. Have your children play with theirs. Don't think in terms of big donations, but just the smallest things that you can do in the here and now. Do whatever you can to deepen the connection.*

At this writing more than a million Smile Cards have been shipped. CharityFocus operates with thousands of volunteers around the world. Three hundred thousand subscribers have joined as members, and its various websites enjoy millions of hits a year. Mehta inspires people around the world in different ways. One family in the Bay Area decided to demonstrate the joy of giving to their son by having his friends tag people with free car washes on his birthday, then taking the children to Baskin Robbins so that they could buy ice cream for everyone waiting in line. Trishna Shah's own industrialist father emigrated with his family from Kenya to America so that the children would enjoy the fruits of an elite American education. Initially aghast that she would accept an internship with CharityFocus when all of her friends had formal internships in for-profit businesses, he grew so impressed by the organization's activities and his daughter's inner transformation that he eventually left his real estate business and took up work in a cancer hospital.

Touched by Mehta's message, Marie, an employee at a large software company, had an epiphany one day at her company's vending machine.[1] She decided that every time she came for her afternoon Coke, she'd leave money in the machine for the next person, with a note and a Smile Card: *Your can of Coke has been paid for. Take this Smile Card and pay it forward.*

From the moment Marie began her campaign, frantic emails began circulating around the office in an attempt to pinpoint the identity of the company's secret Santa. A Neighborhood Watch scheme was set up with two or three employees on constant lookout. At this point, Marie decided that it was time to escalate operations. She moved to another floor, where she surreptitiously left a daily supply of donuts. "For months everyone was talking about it," says Mehta. "It completely changed the dinner conversation."

A third-grade teacher distributed Smile Cards to her class and told them to carry out a random act of kindness that evening as a homework assignment. One young boy, at a loss for what to do, wandered around his neighborhood until he came across a lost dog. He found his owner's address on the collar and led him home, where the owner was effusively grateful. "Hey, no problem," the eight-year-old replied, pulling out a Smile Card. "Here you go—pay it forward."

Mehta's operation manager Trishna Shah, an ex-pat American living in London, has a weakness for Taco Bell's hot sauce, which she cannot get where she lives. Not long ago, when she was in the hospital for a minor procedure, Mehta called upon his American volunteers, and before long, packets of Taco Bell hot sauce began showing up every day in her mailbox. He and "posses" of volunteers hold smile banners at busy San Francisco intersections, so that the traffic comes across them at every consecutive red light. At gatherings to mark the 9/11 Walk for Hope, where people tend to be polarized for and against the American military response to the tragedy, his volunteers are instructed to hug both sides. "We show them that it's okay to disagree," he says. "The polarities have to stop in our hearts and minds first."

Nipun Mehta is swimming against the world's tide. At the moment the most infectious of ideas is atomization—breaking down things into their component parts, treating the parts as if they themselves are complete unto themselves, or considering them as the equivalent of something whole. Virtually all developed countries are founded on a culture of individualism and individual gain. For hundreds of years we have followed a false trail of individual satisfaction as our primary motivation, at great cost. As individualism rises, the indices of every major aspect of life satisfaction, from health care and education to life span and urban safety, fall further among every member of the population, rich and poor.

To the extent that we ignore our natural impulses we work against our nature, which responds to the pull of wholeness. With every step away from the Bond, our natural birthright, we take another step toward separation and alienation, and away from what is best and truest about ourselves. We create more economic crises, more political struggle, more conflict, more calamitous ecological disaster. We erect higher and higher walls between ourselves and the rest of the world. In our real-life Public Goods games, we are now all refusing to play.

We stand at the crucial point in our evolution where we must make a choice. We are one of the most important generations in the history of humanity. With all the calamities in our midst, our choices will affect our children and our children's children—and indeed the world for all time. We can continue our process of obsessively looking for smaller and smaller pieces of our universe, ring-fencing people and things according to more tightly defined differences. We can continue to operate against nature and connect less and less with what we regard as other than ourselves. Or we can embrace the opposite impulse: our natural drive to seek wholeness and connection.

There are tiny signs here and there that the game is changing. Nipun Mehta is what Ernst Fehr and other game theorists would consider an instigator, a change agent in an economic game. Herbert Gintis, professor emeritus at the University of Massachusetts, who has worked extensively with Fehr on game theory and the development of strong reciprocity, discovered when modeling the Public Goods game that the culture of a particular community is not set in stone. If a culture of turn-taking falls apart with too many freeloaders, all it requires is a small group of individuals committed to strong reciprocity to "invade" a population of self-interested individuals and turn the entire thing around. "Even if strong reciprocators form a small fraction of the population, at least occasionally they will form a sufficient fraction of a group that cooperation can be maintained in bad times," he says. "Such a group will then outcompete other self-interested groups, and the fraction of strong reciprocators will grow. This will continue until an equilibrium fraction of strong reciprocators is attained."[2]

Gintis is saying that both selfishness and altruism spread easily, but that altruism is the more contagious impulse. As Fehr has discovered and Lindsay Browning proved with game theory, nice-guy strategies are more disposed to get established and can easily invade social structures. Although this contagion spreads more rapidly in a small group, once one group has a small, stable level of cooperation or selfishness, it may spread. By returning to our natural human tendency to pursue the Bond, even a small group like CharityFocus can make giving contagious.

Instigators like Nipun Mehta instinctively understand the power of the Bond—that in their hearts, human beings most want to belong together. They also understand, as Robert Axelrod discovered with the Prisoner's Dilemma tournament, that niceness is the more powerful strategy, and always outperforms selfishness. Selflessness is the most self-serving act of all because it is the position naturally most beneficial to us all. Our daily lives can be lived as a Nash Equilibrium, taking into account both what is good for our world and what is good for ourselves.

Mehta recently set up another local experiment in generosity to see if a business could run entirely on the generosity of its customers. He persuaded Rajen Thappa, the owner of Taste of Himalayas restaurant, a traditional Indian restaurant on Shattuck Avenue, Berkeley's main street, to allow CharityFocus to take over the restaurant every Sunday from 11 a.m. to 3 p.m. CharityFocus would pay the chefs' salaries and the cost of the ingredients, but its volunteers would provide the rest of the workforce of waiters, dishwashers, and cleaning staff.

There are no prices on the one-page Sunday brunch menu; the check at the end reads $0. But a note is attached that says, "Your meal was a gift from someone who came before you. To keep the chain of gifts alive, we invite you to pay it forward for those who dine after you." Diners may pay whatever they wish, and the amounts are kept anonymous. Each week the donations are used to pay the Taste of Himalayas for ingredients and the cooking staff, and any surplus goes to pay for the following week or for other CharityFocus activities. Every Sunday thus far they have received more money than they need.

Nicholas Christakis, the sociologist and network specialist, recently discovered a pay-it-forward phenomenon in networks. The participants were randomly assigned to a sequence of different groups in order to play a series of one-shot Public Goods games with strangers.

This enabled Christakis and his partner, James Fowler, to draw up networks of interactions so that they could explore exactly how the behavior spreads from person to person along the network. They discovered a scientific demonstration of what Mehta has hoped to achieve: giving creates a contagion of giving, a network of pay-it-forward altruism. The actions of participants affected the future interactions of other people along the network. "If Tom is kind to Harry, Harry will be kind to Susan, Susan will be kind to Jane, and Jane will be kind to Peter," writes Christakis. "So, Tom's kindness to Harry is seen in Jane's kindness to Peter, even though Jane and Peter had nothing to do with Tom and Harry and never interacted with them."[3]

All it took was one act of kindness and generosity to spread through multiple periods of play and up to three degrees along the network. "Each additional contribution a [person] made to the public good in the first period of play is tripled over the course of the experiment by other [people] who are directly or indirectly influenced to contribute more as a consequence," Christakis and Fowler write.[4] For every act of kindness or generosity you do for a friend, he pays it forward to his friends and his friends' friends and his friends' friends' friends. Christakis proved that Mehta's instincts were correct: kindness and generosity create a cascade of cooperative behavior, even in the most hardened of hearts.

Over the generations, as modern society has progressively pulled apart, certain far-sighted individualists have acted as change agents for holism. Even during the midst of the Great Depression's banking crisis in 1929, John Spedan Lewis, who became the head of department stores in Britain after the sudden death of his father, believed that the "present state of affairs," by which he meant that outside shareholders were separating the providing of capital from its use, was a "perversion of the proper working of capitalism.

"Capitalism has done enormous good," he wrote. "But the perversion has given us too unstable a society. . . . It is all wrong to have millionaires before you have ceased to have slums."[5]

By nature Lewis was a strong reciprocator. In this statement, extraordinary for its time, he was saying that too many people owning stock in a company, and far removed from its operation, were freeloaders. He did not believe in socialism, but in a brand of capitalism in which a person's financial reward should be commensurate with his contribution. He also understood that lack of fairness in a society was highly toxic to all, rich and poor. "Differences of reward must be large enough to induce people to do their best," he wrote. "But the present differences are far too great."

Lewis came upon the idea of creating a superordinate goal of his business. He turned his department store into a partnership, with every employee a part owner. No matter how menial a person's contribution, each would receive an array of perks, including excellent pension schemes and country-club membership for weekends away. But the most radical idea of all was that profits would be split among the employees. Although the workers would receive differential salaries according to their contribution, to this day each employee, from the lowest shelf-stacker to the chairman, receives the same percentage payout of his own salary as a bonus.

In early March 2010, when Marks & Spencer, one of Britain's major retailers, made profits of just 5 percent, John Lewis distributed £151 million of its 9.7 percent profits, with every one of the chain's 70,000 employees receiving 15 percent of their basic salary, the equivalent of eight weeks' pay. "It was tough," said one employee about the 2009 recession, "but we all pulled together." Lewis understood that he could create a resonance effect among all of his employees if they were all working for the good of the whole. Like the Oxford rowers, the power of working together raised everybody's game.

Dave Barham represents a typical American success story—with a twist. In 1946, Barham set up a hotdog stand on Muscle Beach next to the pier in Santa Monica, California. Besides offering quality hot dogs and fresh lemonade, Barham fervently believed that his employees would connect better with the customers if they were all having fun. His strategy worked. Over the years, Hot Dog on a Stick became a large chain, with 105 locations throughout America's shopping malls. Barham continued to innovate in ways that enhanced the enjoyment factor and spirit of fun for customers and employees; he added a giant inflatable slide to many of the sites, and in the mid-1960s, inspired by the fashion of the time, he designed a staff uniform of hotpants and jockey caps in bold stripes of primary colors, which were set off by the white stands with mirrored walls. To make each employee feel special and to emphasize the fun aspect of the job, each hat was handmade and customized to fit its owner.

Prior to his death in 1991, Barham decided to share his good fortune with the team he had mentored over the years and created a trust enabling the business to be purchased entirely by the staff. Presently, Hotdog on a Stick is the only fast-food chain in the world in which equity and the brand itself are 100 per cent owned by its employees. As with John Lewis, ownership inspires loyalty and results in low turnover—an unheard of situation in a fast-food chain. Afton, for instance, has been with the company a decade. "Going to each of my shifts, I know that I can make a difference in the company and that my equity plan will be affected by all my actions. And the greatest part . . . I work with people who feel the same way," he says.[6] Barham's motto was to make *fun* a superordinate goal—and the atmosphere at HDOS continues to be infectious on both sides of the counter.

In 1969, a small band of inhabitants of Portland, Oregon, completely reversed the tide of urban sprawl and civic ennui in their home city by organizing themselves into a group called the Riverfront for People to protest the widening of the riverside roadway. What they wanted, they said, was less highway and more pedestrian access to the river. After two years of discussions the Riverfront for People prevailed. Harbor Drive was demolished, Tom McColl Waterfront Park was created, and Portland remains a model of accessible and friendly urban life.[7] More important, at a time when social capital is increasingly absent in the United States, Portland's citizenry continue to be the most activist in the country. They have learned the power of becoming a superorganism.

In the UK, the Coin Street Community Builders, a group of local activists, joined together to successfully oppose large-scale development plans for expensive high-rise housing in a white working-class district of the South Bank along the riverfront of the Thames in London. After creating a center for burgeoning businesses, the community group built state-of-the art public housing and a children's center by donating some of the profits from the businesses they'd helped to create, including a luxury Harvey Nichols restaurant on the top floor of the Oxo Tower. In this way, the Coin Street group made use of community private capital as a virtuous circle: to fund public services to the less fortunate without the need for a government handout.

There are many other examples every day of the power of a superordinate goal. Innovative organizations around America, for instance, have used the power of the Bond to create collective ways of providing public services. Southern Maryland Electric is a consumer-owned cooperative that provides power to a portion of the state, and Group Health Cooperative, based in Seattle, offers members the ability to govern their own nonprofit healthcare system. Realizing the need to work together globally, some religions are even reaching across the aisle. In 2007, a group of Muslim religious leaders published an open letter, "A Common Word between Us and You" to encourage peace and understanding between the Christian and the Islamic faiths by emphasizing the common connection between them; both religions share the religious commandments to love God and one's neighbor.

Nevertheless, in the midst of the continuing crises in our current way of life, we need to take measures far more radical than individual instances of benign capitalism or civic recovery. We now require nothing less than a revolution in our thinking. We must abandon the path of atomization in the way we relate to people, in the way we create our neighborhoods, and, most important, in the way we view our world. It is time for all of us to embrace the new scientific discoveries across all the disciplines, which are proving that we have maintained a false and dangerous view of ourselves. It's time to begin a new Age of Enlightenment, one that recognizes and honors wholeness, and abhors the polarization of one people, one religion or even one political party against another. Like Nipun Mehta, we must all begin our individual efforts to make the Bond our superordinate goal.

In 2009 Karma Kitchen opened its second restaurant, at the Polo India Club in Washington, D.C., with a third added at Klay Oven in Chicago in 2010. Not long after the Washington site first opened, a family visiting from Mexico walked into the restaurant with misgivings. The father, an economist, was highly skeptical about the concept of paying it forward; *if you trust people, they will take you for a ride*. At the end of the meal, while he was debating whether to leave without paying anything, his eleven-year-old son pulled out a twenty-dollar bill—his allowance for an entire month.

"What's this?" the father asked.

"My contribution," he said in Spanish. Despite his father's protests, the boy insisted on leaving the money.

The father looked away for a long time, and then reached for his wallet. He began stuffing the bill holder with all the twenty-dollar bills he had. Before leaving he scribbled out a note: "Today, my son taught me about generosity." After a single act of giving, a potential freeloader became one of the most generous players of the Public Goods game.

CharityFocus is an example of creating a superordinate goal as a life purpose. Its aims resonate deeply with its customers. Thus far both restaurants have received more than they would if they charged for meals. Volunteers travel from all across the world to help out as staff, just to experience service with no strings attached. After basking in the experience of an unconditional gift, many diners are moved to tears. They hug the other diners and leave notes and poems, as well as money. Almost everyone asks of the volunteers, *What can I do? How can I help?* But it's the look on their faces that speaks the loudest. It's as though they have just been reminded of a favorite, half-forgotten tune. Yes, this is it, they seem to be saying: this is what it means to be a human being.

PART IV

TOOLS FOR A NEW WORLD

Most of the feeling we call religious . . . is really the under-standing and the attempt to say that man is related to the whole thing, related inextricably to all reality, known and unknowable . . . the knowledge that all things are one thing and that one thing is all things. . . . It is advisable to look from the tide pool to the stars and then back to the tide pool again.

John Steinbeck, *The Log from the Sea of Cortez*

CHAPTER 13

FIXING YOUR FALLING-DOWN WORLD

The purpose of the first three sections of *The Bond* has been to *change our story*—to show how all of life has been designed for connection, not competition, and to explore how this new story of who we are changes how we should relate to others in our individual relationships, our communities, and our institutions, such as our schools and our workplaces.

Part IV offers you tools to bring these principles into your life. I hope to help you change the competitive and individualistic mind-set that too often forms the undercurrent of our lives and to foster your own deep-seated impulses toward wholeness, cooperation, fairness, generosity, mutual support, and community.

The key to achieving this rebirth, to improving your relationships, your neighborhood, your town or city, and even to successfully tackling the many crises now facing our country lies in reframing the core belief that underpins every aspect of our society: *I win, you lose.*

As I hope I have convinced you throughout this book, this mistaken belief—that *winning equals winning over someone else*—is the greatest cancer of our time, accounting for every single one of the crises we face in the West. Other than on the sports field, competition has been shown to be a dangerous and outdated tool in every regard: in business, education, problem solving, and relationship and even community building. Indeed the competitive mind-set is probably the greatest impediment to progress. The latest research shows that students, employees, managers, business owners, couples, and neighbors are happier, healthier, and far more productive when they work together in collaborative ways.

Schools that use cooperative learning, where A students work side by side with C students, produce better results than do those schools that group children by ability and have them compete for grades or constantly strive to improve their own personal best.[1] Ethnic and political groups work better when they don't try to work only for compromise or to score debating points, but instead come together and speak deeply about their own values, at which point connections and solutions naturally evolve.[2] Collaborative solutions at work consistently outperform those in which companies rate individual performances against each other.

As a success coach hired by Microsoft, the bestselling author and *Chicken Soup* franchise founder, Jack Canfield, observed that Microsoft's practice of creating small internal competitive "silos," vying against each other for performance and rewarded or punished for their individual success or failure, created such a climate of fear that it actually hampered innovation and lost the company market share. This company policy stood in stark contrast to the corporate climate at Google, one of its major competitors, where individuals were encouraged to work as a team, were offered the time and space for collective brainstorming, and were rewarded for the entire group's effort. By removing a culture of naked-claw competition, Google made use of the Bond—and not only created happier employees but ultimately began to produce better results than its rival.[3]

Since the release of the hardcover edition of *The Bond*, many American readers have asked: *But what about the rights of the individual?* Isn't enlightened self-interest the chief driving force of business, education, sport—*everything?* And isn't every better mousetrap that gets built the result of the inventor following his individual dream? How will we ever achieve anything significant—or win at anything—if we don't focus on Number One?

The change of emphasis in our relationships and our society from "me" to "we" will not erode individual rights, ability, achievement, freedom of expression, or ownership in any way. Nor will it require that we relinquish our hard-earned cash or possessions, repudiate our economic system, or overturn our democratic way of life. The only practice we will give up is the need to strive for individual achievement *at another person's expense.* That mind-set in itself is flagrantly anti-individual and undemocratic: somebody's individual rights always get trampled in *I win, you lose* scenarios.

Furthermore, considering the group enables us to perform better individually in all of our endeavors. This was demonstrated in research on sports performance carried out by Veronica Son and Deborah Feltz of the Department of Kinesiology at Michigan State University, who designed an ingenious study on "self-talk," the internal monologue people use to psych themselves up before a performance.

This brand of inner cheerleading is practiced by all sports performers, and thanks to the work of people like metaphysical writer and publisher Louise Hay, whose bestselling book *Heal Your Body* has taught millions to make use of "affirmations" about themselves to boost confidence or effectiveness in many areas of their lives. It works (and there's been a good deal of research into it, detailed in my book *The Intention Experiment*) because our brains, clever as they are in most other regards, cannot tell the difference between an action and a thought.

Most of the research on self-talk and mental rehearsal has focused on internal pep talks about "I" in order to build up personal confidence. Son and Feltz decided to test what happens to individual performance when participants concentrate their self-talk on the group's performance as a whole. They randomly assigned eighty participants in a dart-throwing contest to one of three groups. The first group used self-talk statements that focused on each individual's own ability and performance; the second used internal conversation that emphasized the group's capabilities and performance; and the third, the controls, simply thought neutral internal statements. When Son and Feltz tallied the results, both personal confidence and performance were highest in those focusing on group affirmations. In other words, those using group-oriented self-talk displayed more confidence in the team *but also performed better as individuals.*

This study has enormous implications not merely for sport but for every aspect of our lives because it shows that focusing on the group's efforts naturally raises everyone's game. *Just thinking "we" helps "I" do better.*

Nevertheless, even if you believe every word of this book, changing your belief system can be a tall order for one simple reason: your own internal hard drive was programmed long ago to think individualistically and competitively. *Eat or be eaten. My way or the highway.* That mind-set creeps into both your worldview and your dealings with others in ways that you cannot even imagine, much less identify.

Even when we attempt to create new prescriptions for living, many are unconsciously assembled with some variation of *I win, you lose*. As Einstein once noted, it is virtually impossible to change a paradigm when making use of the same mind-set that created the very structures you're trying to transcend.

At the moment, our paradigm for living is largely the following:

- Life is a zero-sum game (I win only if you lose).
- What's in it for me?
- I must win, dominate, or be first in order to feel good about myself.
- I take care of my own. Otherwise it's every man for himself.
- It's Us versus Them.
- I do best for society by looking after Number One.

If we are to prosper, individually and collectively, each of us must wipe clean our mental hard drives of the sense of scarcity, lack, competition, and extreme individualism with which we are now programmed. To do this we have to challenge the very assumptions and thought processes on which those concepts and assumptions are based.

A good starting place for a new Bond paradigm is to replace the Adam Smith model of "we do best for society by looking out for Number One" with the Nash Equilibrium, as discussed in chapter 8. Our own best response in any situation is to *choose what is best not simply for ourselves but also for the rest of the group*.

Here are the main principles of the Bond paradigm:

- I win *when and only when* you and I win.
- I feel best when I connect, whatever it takes.
- How can I serve you?
- We're all in this together, and we pull together through thick and thin.
- It's Us *plus* Them.
- We do best for society by looking out for ourselves *and* the rest of our community.

Adopting these aspects of the Nash Equilibrium in your dealings with others will encourage you to overcome your internal "I win, you lose" programming and become a spiritual change agent, at home and at work.

In the following pages, you'll find a program of simple practices that will help you to become more conscious of the essential Bond that unites us all and learn to make use of it in one-on-one and group relationships. These tools focus on the four areas highlighted earlier in *The Bond*:

- Seeing the whole
- Changing the way we relate to people
- Enlarging our experience of community and learning to come together for common goals
- Becoming a spiritual change agent through daily acts of generosity and cooperation

These simple exercises will train you to see the world from a more holistic perspective, enjoy more cooperative relationships—even across the deepest divides—develop more united social groups, become a highly infectious spiritual activist for cooperation and connection in your immediate workplace and community, and make use of group support as a powerful local and global agent of change. Practice these activities every day and you will quickly learn to identify when you are operating according to the old paradigm of exclusion, fear, lack, and competition. They will enable you to embrace more inclusive ways of being.

You'll find suggestions about how to use the power of a common overarching goal to help a disparate and possibly disputing group of people put aside individual differences to work together cooperatively. These techniques can turn small groups into a superorganism, quickly transforming a group of strangers into a potent healing circle within their community.

In chapter 14, I offer advice and guidance about how to turn any group into a "Bond pod," a group that works together cooperatively for workplace or community change. By "group," I mean any collection of individuals besides yourself and your immediate nuclear family: your neighbors, your closest friends, your work colleagues, an existing group such as your church congregation, or even members of your larger community. I also offer discussion points, weekly exercises, and challenges for the group that will help to rebuild fragmented neighborhoods. In this way, individual groups can act as ambassadors of connection and renewal for an entire community.

Once you begin practicing these exercises and after you've formed your "pod," you will see that you can help to fix the problems we now face. By a simple change of perspective, and with these individual and collective practices, each of us can transform our culture and move out of crisis, individually and collectively.

SEEING THE WHOLE

The purpose of this section is to help you to develop "aerial vision" in your relations with other people. This includes fine-tuning your ability to notice details about others, particularly people who are different from you. Over time you will learn to question your automatic assumptions, hold opposing ideas in your head without being judgmental, detach from your own prejudices, and refrain from the kind of thinking that traps you into an "Us versus Them" mind-set. Practicing these exercises every day will also sensitize you to the emotional undercurrents of relationships and increase your capacity for empathy.

The first step in the art of seeing holistically is simply learning to notice more. As discussed in chapter 9, the scientific evidence shows that our well-developed forebrains inhibit our ability to retrieve information. The forebrain sees a portion of something (a person, concept, or behavior), labels it with a name, and, in a sense, fills in the conceptual details to produce the whole. Studies show that language often causes "verbal overshadowing," suppressing visual perception or memory. Like animals, we need to learn how to become cognitive "splitters," not "lumpers," and to become sensitized to the smallest detail. You can best achieve this by cultivating your ability to tune into your sensory experience of the present moment and developing areas of awareness that lie beyond language. Carrying out tasks that engage other parts of the brain besides those involved in verbal ability enables you to notice more about the connections between things. After practicing this kind of "aerial" perception, you are more likely to appreciate that in any circumstance, there are always multiple versions of "the truth."

Step 1: Noticing More Detail

For this and many of the other exercises in this chapter you will need to keep a journal in which to write down exercises and keep track of progress.

- *Practice doing nonverbal tasks*—working a puzzle, listening to music, or even drawing.
- *Each day, spend ten minutes studying something*—a building, tree, animal, piece of fruit—in exquisite detail, noting the parts rather than the whole.
- *Work on examining things in "pieces"*—the smells, sounds, look, feel, and taste of the component parts.
- *Study the particulars.* In any given situation, imagine you are a detective and that you have to memorize the room. Examine every detail: the environment; the particulars of scenery or furniture; if indoors, the people or animals. Resist your mind's tendency to fill in the details of an experience to create a totality. Zero in on specific things—a vase in the room, a picture on the wall—and study every section of it. Experience these component parts.
- *"Listen" with your five senses.* Pay attention to the smell, taste, and kinesthetic feel of a situation, not just the visual and auditory information.
- *Practice this acute sensory experience in ordinary situations.* Four or five times a day, when, say, preparing dinner or brushing your teeth, stop thinking thoughts and tune into your five senses. What do your present circumstances smell, taste, and feel like, as well as look and sound like? As noted in chapter 9, practicing this kind of "mindful" attention numerous times a day will enable you to notice much more detail within a few weeks.
- *Study your partner and children, pets, friends, and work colleagues.* Observe every part of them closely during a variety of activities— without judging them negatively or positively.

- *Take time* to study landmarks on your way to and from work or school. Notice them from all angles and also the relationship between them. To navigate through complex landscapes and waters, indigenous tribesmen like those mentioned in chapter 9 develop keen observation of detail around them.
- *Become a student of your local weather and (if nearby) water.* Observe subtle changes in the wind's direction and the different patterns this causes in the water.
- *Observe the response of birds and other animals to weather.* If they seem to demonstrate strange behavior, note whether you have any feelings yourself, such as a headache or other physical symptoms.

Step 2: Study Your Own Thinking Processes

- *Record in your journal your main thoughts* every day for several weeks.
- *Become more mindful of what you are thinking most of the time.* Become a mental note-taker, acknowledging what is rather than judging what should be.
- *Work on staying neutral and observing without being judgmental.* A rainy day isn't a bad day unless you think it is.
- *Accept all that happens.* This means putting away all opinions and interpretations of what goes on. Stop yourself from clinging to certain views, thoughts, opinions, and preferences and from rejecting others. Accept your own feelings and experiences, even the unpleasant ones.
- *Whenever you realize that you are judging what you see, think "I am thinking" and return to observing with simple attention.*
- *Catch yourself when you are making a negative judgment about a loved one.* Turn the negative judgment into a positive loving thought or a sense of compassion.
- *Become conscious of your feelings toward your enemies as well as toward your friends.* When you see someone you don't like, observe yourself as you make internal negative judgments about that person. Pull yourself back into the present encounter.
- *If you catch yourself being negative about anything, take your mind back to a positive situation you've experienced, or simply stop and become mindful of your breath.*

- *Start to recognize a pattern.* When do you have the most negative thoughts: at work, in your relationships, around certain people? When are you at your most judgmental?
- *Work on considering life from the point of view of your loved ones or those whom you encounter in your daily life.*
- *Observe how changing your view of others changes your day.*
- *As you begin to monitor your thoughts and record them in your journal, see if your neutral or positive thoughts begin to outweigh the negative ones.*
- *Try not to try.* Work on stopping your expectations or striving for certain results.

This new way of thinking may appear at first to entail an involved analysis of every one of your actions, but after a while you will become accustomed to this larger way of looking at life and at your own actions.

Step 3: Seeing the Unseen

To see the whole, you need to become exquisitely sensitive to your feelings and sensations and the information you are receiving without your conscious knowledge. This includes picking up information beyond the senses, which all of us can do once we become sensitive to our own feelings and trust our intuition. Here are some practices that will help:

- *Check your feelings in every situation.* Are you feeling especially good or bad about a situation? At ease or uncomfortable? Is it the people you're with or the situation itself? To find out more about what most affects you, mentally divide the situation into component parts and test your feelings about each component. Ask yourself in each instance: How does X make me feel?
- *Cultivate your own natural sensitivity to others.* Slow down and learn how to listen to others and to feel their feelings.
- *Read body language.* In conversations, pay attention to what people are trying to tell you without words, particularly by the way they hold their bodies.
- *Study your partner, friends, children, and colleagues again, but this time see what thoughts you can pick up without using words.*

- *Pay attention to intuitive communication.* When you have a gut hunch or an internal message pops into your head, respect and act on it.
- *Practice intuiting an answer.* A tried-and-true means of improving your intuition is to practice guessing with a pack of cards. Keep the deck face down. Concentrate on each card as you guess what it is. Name the first card that comes into your head. If you wish, you can start off with just ten cards so you have fewer choices.
- *Cultivate the art of listening to all sounds in your life:* the rumble of a passing truck, the barking of a dog, an airplane flying overhead.
- *Draw your dream images without trying to analyze them or figure out what they "mean."*

Step 4: Strengthening Your Gut Hunches

Developing your intuition can be a learning process. It's a matter of honoring the still, small voice inside yourself that says "Don't do that!" or "Go ahead and take a chance." You can teach yourself to listen to your gut hunches by practicing the following:

- *Slow down.* We rush about so much, using cognitive reasoning to steer us around, that we often don't have time to listen to our inner voice. When making a decision, mull it over in your mind and wait to see what comes up.
- *Focus on "low-road" information.* As we covered in chapter 9, too often we dismiss the immediate (and usually correct) information while our analytical "high-road" brain tries to make sense of it. Our first, intuitive sense about something or someone is invariably correct.
- *Receive information through your senses first.* In any given situation pay attention to the first image (or sound or feeling) that pops into your head. The famed psychic Ingo Swann, who is gifted at remote viewing—the ability to perceive objects or events beyond normal human vision—tells his students that there are different stages during intuitive receiving: at first, you become aware of simple outlines of shapes, and after a while you can sense more detail. If need be, draw this information before you name it.

- *Slowly fill in the details.* After a few moments, redraw the information you've received, if only mentally (by then you should have received more information). A few seconds later, draw it again. Later you can even try constructing it in three dimensions, with clay.
- *Resist analysis.* When you get an intuitive sense about something, resist the urge to figure out what it "means." This kind of "cognitive overlay" often distorts the truth. I once attended a conference in which the speaker invited us all to remote-view the object hidden inside a giftwrapped box. The first thing that popped into my head was the word "candy," but I immediately dismissed the thought as irrelevant because I live in the United Kingdom, where candy is referred to as "sweets." I then clearly saw in my mind's eye the image of an oval, and because I was wearing an oval brooch, my analytical mind concluded that the hidden object must be a pin. In fact, my intuition, unfettered by this cognitive overlay, had been right all along. When opened, the box contained oval-shaped chocolates.
- *Follow your gut hunches, no matter how irrational they are.* If you get a strong impression that you or someone else shouldn't do something, listen to it.
- *Write down in your diary any premonitions you may have about changes in the weather for the worse.*
- *Over the next year, compare your premonitions with the resulting weather, and note any correspondence between the two.*

Step 5: Seek Out the New

Brain research by Richard Davidson, a neuroscientist at the University of Wisconsin, confirms that the brain is highly malleable and that sections change or grow according to how often they are activated. For instance, meditators who spend a great deal of time focused on a mantra or on their breathing can expand that portion of the brain dealing with highly focused attention.

If you work on brain areas that heighten your curiosity, you will reinforce and strengthen the neural "seeking" circuitry, as identified by Jaak Panksepp, the psychologist at Bowling Green University mentioned in chapter 9. This will help you to stay alert to the details of your environment and also enhance your intuitive ability.

To expand your natural sense of curiosity (and that part of your brain):

- *Engage in frequent "hunting."* Go to flea markets, garage sales, auctions, and trade fairs, where buying activity encourages you to "snoop" around.
- *Solve problems for fun.* Crossword puzzles, jigsaw puzzles, Sudoku, and the like all engage the problem-solving portion of the brain.
- *Watch or read whodunits or thrillers.* Perhaps more than any other form of storytelling, the thriller focuses entirely on "what happens next" or who is responsible.
- *Keep abreast of the new.* Curiosity is essentially fascination with novelty of any sort, even an old story with a new twist. It doesn't have to involve scenarios or information that is completely new. Stay up-to-date about what is happening on the political and cultural scene.
- *Avoid routine.* Study a new subject, take a new course, try new recipes, go to new shows, take a new route to work, go to lunch with a new friend, or put up a new curtain in your living room.
- *Investigate a new area of your town.* Look at details. Actively dig out information about its history, buildings, occupants, and businesses.
- *Hunt for answers.* Set yourself a new intellectual problem and seek out the answer to it.
- *Be nosy.* If you're curious about something, don't dismiss it as unimportant. Track down information or find out more.
- *Learn to listen.* Be genuinely curious about colleagues and friends. Ask open questions about their lives and stay present when they answer you.
- *Be curious about the big questions.* Do some scientific or philosophical searching into the meaning of life and other philosophical issues.
- *Revel in the anticipation of things.* If something good happens, notice how excited you were beforehand. Tease out exactly how you were feeling and what exactly you were most excited about. When that feeling comes again, stay present. Those are the moments when your antennae are most sensitive and you become most aware of everything around you.
- *Notice when you are apprehensive.* Learn to recognize when you are feeling wary of something. What does it feel like? Determine

whether you were right to suspect a certain situation. When that feeling comes again, listen to it.

Step 6: Seeing the Whole Truth

Now that you have practiced noticing more details in your life, you can begin to apply these new skills to your perception of other people, particularly those who are not like you. When you relate to others, do the opposite of what you've learned in critical thinking or debating classes. Focus on what is positive, truthful, and wise about what others are saying, not the flaws in their arguments. Look for areas of common ground.

- *Keep alert for "Us versus Them" thinking, language, and actions.* As soon as you start generalizing about one race or ethnic group of people—whether Republicans, Muslims, or even bankers—you have defined a group as "them." Expunge this kind of language from your vocabulary.
- *Practice maintaining respect for a view that is different from your own.* There is no right or wrong perception of the world.
- *Question anything you regard as a division between yourself and others.* This includes your automatic assumptions about neighbors you don't know, acquaintances of a different ethnicity or religion, the countries and people beyond your borders.
- *Separate out gradations of belief.* The idea that all the people who hold a certain view have the same exact position is a stereotype. As Don Beck described in chapter 9, fine gradations of belief exist among people who appear to agree, and most of us outside of a belief system don't appreciate the wide spectrum of beliefs held within any given position. Pro-choice advocates, for instance, have widely divergent views from each other, from those who believe that abortion is justified in all circumstances to those who believe abortion is justified only in cases of rape. Seek to identify these fine distinctions in beliefs or practices so that you do not miss a chance to find common ground.
- *Look upon everyone involved as a potential partner and honor the need to connect with them.*

- *Seek out kernels of truth in any opposing position rather than the differences in views between the other person and yourself.*
- *Mentally swap roles with someone else.* Imagine the issue from the polar opposite of your own position and offer as many solid arguments in favor of that position as you can. This helps you to take a larger perspective on the issue. By the same token, try to imagine someone else looking at your ideas. What do you think they see?
- *Don't be afraid if you disagree.* Explore the values you have in common and seek a creative solution (and see page 234).
- *Experiment with a positive description of your differences.* Instead of complaining, "Those atheist heathens don't go to church," think and say, "How interesting—atheists believe in a natural intelligence."
- *Name your feelings.* Then reflect back what you believe to be someone else's feelings.
- *Be flexible.* Abandon your entrenched position if it no longer serves.

Practice Exercise: Create Your Own Web of Connection

Examine the various people in any group to which you belong (including work). Select a contentious issue, such as abortion, gun control, taxes, or a local issue at home or work. Draw a mind map of all the different opinions or views on the subject. Then identify the connections between those positions. What interests or values do some of the different positions share? Draw a line between positions that share a common value. Notice how your map forms an interconnected whole.

RELATING IN WHOLENESS

Currently most relationships are forged from the erroneous idea that we have to be the same to get along and that differences between us are to be avoided at all costs. In fact, conflict is considered so antithetical to the human experience that when others disagree with us, we conclude that they must be stupid or ill informed. To justify this position, we find it necessary to debate them, demonize them, and announce their ignorance to the world. In our minds, conflict can be resolved only with *I win, you lose.*

The key to a more holistic relationship with *anyone* is to conceive of the relationship as a "thing in itself" and to focus on the "space in between"—the glue that holds it together—especially when you do not agree with each other. Look upon conflict or differences of opinion as an opportunity to create something new together that you would never generate if you both agreed with each other. It's also an opportunity to forge a new relationship.

Once you view yourself as part of a bigger whole, you will begin to act differently toward others. By making this one simple change of perspective and offering yourself as a vehicle of service to the connection, you will easily find the Bond that is always present and embrace difference within that larger experience of connection.

In the following practices you will explore techniques of relating that allow you to make deep connections with anyone, even those who disagree with everything you stand for, and engage conflict in creative ways in order to produce greater shared understanding and possibility. You'll also learn how to promote closeness through the power of deep truth and candid disclosure rather than the strength of your arguments. During this type of deep sharing, the pull of wholeness builds trust and loosens attachment to fixed positions.

Step 1: Change Your Perspective about Relationships

The first step is to change your perspective about what relationships are for and to lay down some new rules of engagement:

- *Consider yourself a vehicle of service to the connection between yourself and anyone else—no matter how different—and make that your goal, not a self-serving end like being right or getting something from them.* Move beyond your comfort zone and consciously connect with people who are not like you.
- *Open yourself to the highest good in the other person and vow to trust him or her at face value.*
- *Listen actively—with your heart, mind, and soul.* Listen to other things besides what the speaker is saying: how he describes things, what he emphasizes or where he places the most energy, how he holds his body, how he appears to feel.

- *Catch yourself when you are judging someone and examine your emotions when you do.*
- *Focus most on creating a relationship with the person who is most in opposition to you.*
- *Uncover the hidden connections between you and other people, whether in your faith, your locality, your citizenship, your sex, or your local or national interests.* Both Republicans and Democrats have many identical interests: a love of family, God, children, home, and country. All of us want to fix the economy, the roads, the government, the high price of gasoline, our educational system. Working out how to do so together affords us an opportunity to come together for a larger goal, at which point superficial differences diminish in importance.
- *Look upon those with opposing viewpoints as "we" who are attempting to solve a problem together through a "third way."*
- *Share honestly and reveal the deepest aspects of yourself when you interact, rather than engaging in superficial chitchat.* Encourage others to do the same.
- *Observe how a different perception of humanity develops when you and the other person share deeply.*
- *Allow different realities simply to coexist—again without being judgmental.*
- *Engage in differences creatively for greater possibilities.* (See page 234 for more specific recommendations.)
- *During differences, seek to change the space between you.* Change the energy that flows between you by changing your tempo, your attitude, your facial expressions. Make your body language and unspoken communication transmit your desire to connect.

Step 2: Learn the Art of Compassionate Meditation

The most important way to move beyond the boundaries of our own separateness is cultivating nonspecific altruistic love by practicing the compassionate meditation technique discussed in chapter 10.

Practice thinking daily: "I appreciate the kindnesses and love of all living creatures. May they be well and free from suffering." Think this thought first about your closest family or relations, then your friends, then your acquaintances, and finally your enemies. Practicing this meditation daily for just a few weeks will promote a sense of greater compassion within you for all people.

Step 3: Use "We" Affirmations

Many of us now realize that the kinds of thoughts we think throughout most of the day create our reality, and we try to program ourselves with positive self-talk, such as "I am going to have a positive day. All good things will flow to me, and everything that's supposed to happen will happen."

Instead of thinking or repeating positive affirmations about yourself, try to frame the same thoughts about your immediate group as well as yourself. Suppose you'd like to carry out an affirmation about doing well at work. Instead of focusing on "I," think of the team you work with and say out loud or in your head, "We all do well in our job today." If you are using an affirmation about your own health, think of your entire family and say to yourself, "My family is always healthy." Thinking of "we" will help "I" do better.

Here are some other common affirmations and how to change them to "we":

- I have all that I need.

  We have all that we need.

- I am cherished and loved.

  We are cherished and loved.

- Things are getting better and better.

  Things are getting better and better for our neighborhood / all of us / our group at work / our company.

- I know where I am headed.

  We know where we are headed.

- I'm going to succeed.

  We are going to succeed.

- I am healthy and well in every way.

  We are healthy and well in every way.

- I trust that the universal mind will send me the answer.

  We trust that the universal mind will send us the answer.

In your journal, list your typical thoughts about the future and write underneath each one a new positive alternative: where you'd like to end up. Then write the same forecast, but about all of the people involved in that event: your family, your friends, your neighborhood, your workmates.

Step 4: Cultivate the Art of Bonded Conversation

- *Attempt deep understanding.* Reflect back what you think you heard.
- *Focus on observations, not your feelings or opinions.*
- *Ask questions to clarify your understanding, not to score points, and use language that empowers the other.* Instead of "What are you doing about X?," which sounds accusatory (as if they aren't doing enough), say, "What would you like to see done about X?"
- *Never interrogate.*
- *Reveal how you feel about something.* Get in touch with your real feelings, and be sure to avoid evaluations or criticism of the other. Stick with your feelings, not your judgments. ("When you do X, it makes me feel Y.")
- *Express your needs truthfully.*
- *Request what you'd like to have happen that would best meet your needs.*
- *Listen when the other person in your conversation does the same.*

Ten Steps to Overcoming Deep Conversational Divides

Create a safe environment. Get the other party (or parties) to agree about what subjects are comfortable to discuss and what conditions make it comfortable. Then lay down ground rules enforcing these conditions.

1. *Invite cooperation from the beginning when raising an issue.*
2. *Listen to the other with a view to understanding, not agreement or disagreement.* Don't try to win the debate, convince others of your "rightness," or convert them to another point of view.
3. *Stay absolutely present, as you learned to do when practicing mindfulness.* Don't let your mind wander or begin formulating your response when the other person is speaking.
4. *Identify and explore the other person's core values and interests:* underlying hopes, needs, values, concerns, motivations, fears, and ideals. Explore all interests that are mutual.
5. *Separate the other person's suggested solutions for satisfying the core interests from the interests themselves.*
6. *Reveal the backstory of why you believe what you believe.* Sharing deeply invites deep understanding and connection.
7. *Slow down before responding.*
8. *When disagreements occur, listen harder.* You'll often find that the problem is a gap in your knowledge of the events in the person's life that have led him to that position.
9. *Brainstorm creatively together all possible ways to solve the problem.* Imagine a positive outcome.
10. *Work together on an agreement that meets both your interests as much as possible.*

A LARGER VIEW OF "WE"

You are now ready to apply the principles of the Bond to groups at work and in your community. These exercises and the suggested activities that follow will reveal the power of a larger, communal ("superordinate") goal to transform the atmosphere in your neighborhood or office from "me against them" to "all of us pulling together." As we explored in chapter 11, the scientific evidence shows that people who fire together wire together; whenever a group works together for a common goal, the brains of all parties begin to get on the same wavelength, strengthening the Bond within the group. Coming together in small groups with a superordinate goal provides a social cohesion beyond money, job, or size of property. A common and larger purpose creates instant closeness in any social setting and offers an excellent tool for maintaining cooperation in your office or neighborhood.

Furthermore, as many experts in conflict resolution recognize, working on a shared problem helps to unite people who are in opposition on other issues. For instance, the Search for Common Ground project managed to connect two sworn enemies, the Macedonian Slavs and Albanians, by encouraging them to work together on a project to clean up their local environment.

Communicating with Dialogue, Not Debate

The best place to start is to lay down a new means of group communication. In *dialogue,* as opposed to discussion, a group explores feelings and ideas in an unstructured way to create greater understanding, deeper connection, and a new synergy of ideas.

When attempting to talk together about subjects that matter most to us, we usually speak from our own version of the truth and *invariably end up disagreeing with anyone whose version differs from ours.* Dialogue is a method of discourse that slows down the process of conversation so that your own prejudices get revealed and new possibilities can unfold.

When you are working with a group, especially when discussing any change with people of differing opinions, observe the following practices.

THE RULES OF DIALOGUE

If you are meeting with more than two people, form a circle with your chairs before you begin. It's often helpful to elect one person to be the mediator, to remind participants, who are often schooled in the art of debate and competitive point scoring, about which conversational tools help maintain supportive dialogue. The mediator should be alert to potential arguments, name-calling, stereotypical behavior, unfairness, or disintegration of the process. He or she should also keep time and ensure that the conversation stays on point.

- *Build trust by sharing goals and reasons for participating.*
- *Frame the dialogue as a series of questions, not topics to be debated or discussed.* Plan questions in advance.

- *Start out with a single question.* Give everyone a chance to answer.
- *Go deep, but don't debate.* The purpose is not to reach a decision or have a debate, but to explore and share deeply.
- *Don't monologue.* Each person should have an upper limit of time in which to put forward his or her view.
- *Be alert to your own emotional reactions, particularly to those whose views differ from yours.* What does that surge of emotion say about your own views or prejudices?
- *Engage in deep truth telling about what really matters to you in your community—or indeed your country.*
- *Be fully present.* Listen with heart and mind.
- *Don't be judgmental, no matter how different someone's worldview or actions are from yours.* Just describe the actions or view and your response to it. ("When she does X, I feel Y.") Talk only about what you think and feel, not your assumptions about how the other person thinks and feels.
- *Root out misperceptions and stereotypes in your views of others during the process.*
- *Avoid generalities ("always," "never"), and speak just to the facts: what happened in particular circumstances.*
- *Make it personal.* Tell stories from your own past, which helps to place your interests and point of view in context. Speak always in the first person rather than on behalf of a particular point of view. As the Cambridge women discovered (in chapter 10), telling our own stories puts a human face on issues and helps to forge connection. Describe certain key turning points in your life, your heroes, role models, parents, or parent figures. Reveal what you most dream about.
- *Ask open questions unrelated to the area of conflict in order to get to know the person whose views differ from yours.* This helps to solidify trust.
- *Reflect back what you think the other person has said.*
- *Frame questions about a contentious issue in neutral language, without implied judgments.* Instead of asking "Doesn't your health care plan care about the millions of American children not covered by health care insurance?," say something like "What do you believe should be done about children not covered by health insurance?"

- *Look for common interests, emotions, values, or experiences that emerge in the conversation.* Ask the facilitator to point them out. As time goes on, you will be amazed by the similarity between your core values and interests and those you have considered your opponents'.

THIRTEEN COMMUNITY SUPERORDINATE GOALS

Use or adapt these ideas to turn your group, neighborhood, or community into a communal savings bank to ease individual worries, to improve aspects of your community, and to bond in the process.

1. *Form a "landscape brigade."* Take turns having the neighborhood come together to landscape a neighbor's backyard.
2. *Work as a group to "build a barn."* Build together something for a neighbor—a fence, a wall, a bookshelf, a foundation.
3. *Plant together in communal areas.*
4. *Band together in hard times.* Bring food or other types of support to local people who have lost their job or home.
5. *Organize your neighbors to protest as a group against a proposal that will negatively affect your neighborhood.*
6. *Form a Neighborhood Watch to lower crime.* Make sure everyone on the block takes a turn.
7. *Create a neighborhood or community campaign to clean up litter, improve the parks, improve medical services, or lower crime against children.*
8. *Take turns tithing your time to the parks, the hospital, an old people's home, a hospice.*
9. *Set up a neighborhood "savings bank"* (see chapter 11). Gather twelve people to create a communal savings and loan, as with the *moai* in Okinawa. Each party contributes a set amount every month with set interest and takes turns collecting the monthly pot. If you'd rather not use money, substitute food, a household spring cleaning, gardening, growing food, attic cleanups, or the like. Or barter by trading work or products instead of money.
10. *Cook or bake extra and share around the neighborhood.*
11. *Support your local school by taking turns teaching skills to the students.*
12. *Take turns walking each other's dogs or driving the neighborhood children to school.*

13. *Start a local partnership for the community's health care center, utilities, or any other service owned by and run by the community* (see page 212). Or start eco-schemes to save energy and recycle in the community.

BECOMING A GAME-CHANGER

Most people are terrified by the many crises we now face and frustrated by the inability of our leaders to solve them. When confronted with the sheer enormity of the problems now before us in every sector of our lives, we feel unable to fix anything ourselves.

This fear grows out of the mistaken notion that the crises in our midst can only be addressed from the top down. However, as I hope this book has convinced you, the change that is necessary—the one that will truly solve most problems in our individual lives, our society, and indeed our world—is not just a change of policy, a new law, or a tighter regulation of some sort, but a fundamental change of heart.

This type of change must come from the bottom up—from ordinary individuals making changes that ultimately cause a contagion of change in their neighborhoods and workplaces. This change starts with you and me, in the fundamental way that we engage with the world.

It is possible for each one of us to become a "spiritual citizen" and transform our own life's purpose from "me and mine alone" to "ours." Simple everyday acts of generosity will enable you to become a powerhouse of change, permanently shifting the culture of greed and materialism around you, creating trust and setting off a contagion of giving and cooperative behavior.

Become Conscious of Your Emotional Influence

As noted in chapter 10, your thoughts and emotions are highly contagious, and even your attitude to life has a profound effect on the people around you, affecting not only their emotions, but their bodies and their ability to function. Arguments or minor conflicts profoundly affect our immune system, the number of our natural killer cells, cortisol rhythms, and even the function of the hypothalamus-pituitary-adrenal gland axis—all regulators of the body's ability to fight off disease.

To become a spiritual activist, you must first become a positive influence on those around you. Create a "relationship diary" within your journal to track your effect on your partner, parents, children, friends, and coworkers, and become more mindful of your emotional contagion. Writing down the particulars of your interactions with others will enable you to track how contagious your actions are and help you to become a force for good.

- *Consistently engage in activities that promote a positive frame of mind.* Studies have shown that the more we engage in activities that result in positive thoughts, the more positive about life we become. If you are deeply dissatisfied with your current situation at home or at work, focus on changing it.
- *Work on "rekindling" your brain to be positive about your relationship with your world.* Neuroscientists now understand that the brain molds itself to our thoughts. A depressed person who remains untreated becomes resistant to improvement because sadness and negative thinking have become hardwired in the brain.
- *Examine the full repercussions of all your actions.* Who or what do your thoughts and actions affect? What is the ripple effect of your actions or thoughts? Is anyone hurt? Who is most helped by them? Approach all your thoughts or actions from a sense of holism—how they affect the world.
- *Consider whom you might be affecting when you are in a bad mood.* Become mindful of the fact that your anger or sorrow affects everything around you. Begin to consider how you speak about people and things in your environment.
- *Detail times of your emotional highs and lows, and make note of any corresponding effects on your partner, friends, children, or coworkers:* their health, mood, mental and emotional state, and effectiveness in the world.
- *Note any interesting correlations.* When you are angry about something, is a partner, friend, child, or coworker clumsier than usual, or less able to carry out tasks that are ordinarily easy? Also note how your good moods affect them. Do they suddenly become happier and more effective in the world?
- *Monitor your movements.* When you make negative hand gestures,

note any physical, mental, or emotional effects on your loved ones or associates.

- *Search for correlations between your positive and negative facial expressions.* When you smile, what happens to your partner, friend, child, or coworker? What's the effect when you scowl?
- *Examine the emotional state of your relationship around the time of any injuries, no matter how small.* Are you or your partner, friend, child, or coworker healing quickly or slowly? Make a note of your views about the healing rate.
- *Practice switching roles with some of your loved ones.* Imagine what it is like to be your partner or spouse, your parent, your child, your coworker. Get inside their shoes and feel what it would be like to see the world through their eyes, with their hopes, fears, and dreams. Think how you would respond.
- *When someone you know is suffering, imagine what it is like to be him and to be faced with his current crisis.* Try to feel and have empathy for his suffering. Ask yourself how you'd feel if you were suffering in this manner and how you'd most want to be healed.
- *Ask for feedback from your loved ones about your best and worst feelings.*

Ten Acts of Spiritual Activism

1. *Do something for a neighbor you dislike.* Take out his garbage, mow his lawn, look after his children.
2. *Blog a "Daily Good News" and email it to everyone in your office.*
3. *Bake a batch of Cookies of the Week to take into work each Monday.*
4. *Avoid ostentation at home.* A new study shows that the wealthiest people in any neighborhood often distrust their neighbors the most and consequently suffer the most ill health.[4] Why have the most grandiose house on the block if it only causes discomfort to everyone else and makes you sick?
5. *Offer one item of praise to a neighbor or coworker every day.*
6. *When you perform well at work, announce to the boss how well your group did as a whole, rather than scoring points for yourself.* Suggest that the entire office get rewarded for a good performance.
7. *Stop yourself when you are trying to outdo someone else's story about her*

children by telling a story of your own kids' (superior) achievements. Avoid comparisons about anything, but most particularly about your children.

8. *Open your door regularly to your neighbors, particularly those who are new or different.*

9. *Pay it forward.* Pay for someone else's parking meter, a toll, a hamburger, a soft drink, a movie ticket—you name it.

10. *Hug a Republican (or a Democrat) today, whichever one is not a member of your political party.*

THE UNIFYING EFFECT OF FAIRNESS

Adopting fair practices at home and at work quickly unifies people and establishes trust, largely because ideas about fairness appear to be universal. In a recent study, researchers at the Harvard Business School asked Republicans and Democrats to design the ideal society for wealth distribution. *Both came up with a markedly similar picture for a just society.* The result did not bear any resemblance to America, Canada, or even the United Kingdom. The picture drawn by both sides most resembled socialist Sweden, where there is far less division between rich and poor than there is in our country.

Although we may be polarized in many areas, all of us—rich, poor, Democrat, Republican—broadly agree on what is fair.

You can reestablish fairness in your life, in your community, in your workplace, and in your country by following the ten simple Fairness Principles below and in free downloads, which you will find on the Bond website (www.thebond.net). *Feel free to pass them on to all your friends.*

The Ten Fairness Principles

1. *Adopt the golden rule as your daily mantra.* Do unto others as you would have them do unto you. When you are fair, others respond in kind.

2. *Don't be afraid to openly stand for integrity, honesty, trust, and mutual benefit.* Let your actions and not just your words declare your desire for fairness.

3. *Encourage approaches to problem solving that involve the whole group at home, at work, and in your neighborhood.*

4. *Choose solutions that are both good for you and good for every group with which you are involved at home, in your neighborhood and community, and at work.*

5. *When you disagree with someone, seek collaboration rather than a competitive or adversarial approach.*

6. *Support only those businesses, organizations, and policies that place transparency and fairness to the consumer above profits.* Refuse to support any business or policy that engages in corporate cheating or other practices that deliberately harm its competitors.

7. *Support only those organizations or institutions that engage in practices that are not unfair to any sector anywhere in the world.* Likewise, support laws and policies that provide equal opportunity for all citizens of every persuasion.

8. *Refuse to engage in any activity at another person's expense, and stop rewarding practices at work, at home, or in your neighborhood or community that are manifestly unfair or that encourage unfair advantage.*

9. *Encourage fair play and teamwork over winning at all costs in every circumstance, and teach this to your children.* Encourage them to choose words and actions that benefit both themselves and those around them—even those who are not their friends.

10. *Even if life sometimes seems unfair, you don't have to be.* Take a daily fairness inventory of your actions, and drop those that are unfair or unduly competitive.

STAYING BONDED AT WORK

The corporate world is just beginning to see itself in terms of its interdependence in the world: on whom it relies and whom it affects by its actions. The most forward-thinking corporations look at the usual dollars-and-cents issues such as product development and R&D in terms of "value creation"—solutions to global problems that also create corporate wealth. These types of organizations also examine their own corporate social responsibility: who is hurt or helped by what they make or do. By examining the ripple effect of your own company, you can ensure that it acts as a "global steward" that not only makes a profit, but also promotes world unity.

- *Make a heartfelt examination of your company's mission and products.* Whom and how many does it affect? Are they affected for good or ill?
- *Prepare a flow chart that examines all the activities of your company—* from R&D and gathering of raw materials to manufacture of the product, distribution, and sales—and note who or what is affected. Who are you helping? Who are you hurting?
- *What is your long-term vision for your work? What are your values?* Do your company's vision and values match yours? If not, consider changing the company's focus or, if you can't, consider finding employment elsewhere.
- *Ask yourself what your company could do to promote unity in the world.* If the company's products don't help the planet, how might it shift its focus to do so? For instance, if your company handles public relations, why not try to focus on those clients who attempt to have a positive impact on the world?
- *What procedures can you put into place to enhance rather than harm?*
- *If your company's business is causing harm and divisiveness, how can you change the company's mission to promote world unity?*
- *What other companies can you connect with to discuss ways to unite with the rest of the world and enhance the quality of life globally?*
- *If you manufacture abroad, how can you employ local residents fairly in your plan?*
- *Redress the injury to the planet you have already caused.* If your company has engaged in practices that have harmed the environment, what policies of restitution can it adopt? Are there ways to restore the environments you have damaged?
- *Look at the true safety issues of your company's products* (not what's said in the press releases). Be as objective as you can, and don't swallow the corporate line. If your company is testing drugs in the Third World, seek other employment.
- *Reduce the health and safety risks of your company's products.* Carry out a health audit on the technologies and operating procedures used by your company to create your products. What kinds of contingency plans are in place for emergencies?

- *Be honest with the public and your customers.* If your company has produced a product that is dangerous to people or the environment, have you announced this in a timely fashion? What about people living near your facilities? Are you damaging their environment with industrial emissions? Most companies attempt to handle such issues through "damage limitation" in the belief that, if such a mistake gets out, the company will be ruined. However, when McNeil recalled an estimated 31 million bottles of Tylenol because around six bottles had been tampered with, its reputation soared.

Growing Through Sharing

Most organizations cloak their operations in secrecy. The managers of virtually every corporation in the West believe that staying ahead of the corporate game is achieved only by keeping all future business plans, innovation, inventions, and client bases under lock and key. However, the most profitable businesses are beginning to recognize that when we use the Nash Equilibrium inside our businesses and with our competitors, everyone wins more. Here are a few ideas to try:

- *Meet regularly with all the members of your team.* Frequently remind one another of your company's mission statement of service, long-term vision, and values.
- *Work together on corporate superordinate goals.* As noted in chapter 11, working for a common goal helps to grow connections between people.
- *Follow the lead of "b" companies, who look to the benefit they offer to humanity, and not simply profits, as a measure of corporate worth.*
- *Design rewards for all employees, not simply individuals.* If innovative ideas have produced higher profits, consider bonuses for the entire team.
- *Examine ways to make your company a partnership.*
- *Meet with your competitors.* Find out what they need to improve their businesses. Consider elements of your business that can fit their mission, and vice versa. What can you both do together that will enhance each of your businesses in the process?
- *Get used to thinking of your competitor's needs.* When you begin solv-

ing his problems, you often create an opportunity that benefits you as well.

- *Trade customers.* One of the best and fastest ways to grow your own business is to trade lists with others. Take turns mailing information to each other's customers. You'll be surprised at how quickly you both grow.
- *Have a presence on each other's websites.* Create links or advertise on each other's sites. Offer each other a small percentage of new business.
- *Look to create partnerships with companies with a related business.* Meet and brainstorm together.
- *Tithe.* Give away a percentage of your profits to a good cause. That sends out a powerful message to the universe, as well as your own employees or fellow workers, that you're not just in it for the money.
- *Suggest that your business donate a few days of its employees' time to carrying out a collective good deed.* Choose a special charity that is meaningful to your business and let everyone be involved in helping out on a periodic basis. Work in a soup kitchen or visit a nursing home over Christmas and sing Christmas carols. Have a corporate day out where every member of your company works for Habitat for Humanity. Establish a rota for working with disadvantaged youth. Your time is your most precious resource.
- *Trust in the power of the Bond.*

CHAPTER 14

STUDY GUIDE FOR BOOK GROUPS AND COMMUNITY GROUPS

HOW TO BUILD YOUR BOND POD

This study guide focuses on the essential Bond that unites us all as a means of strengthening our connection with other individuals. It aims to empower you with the skills necessary to enable individual groups to become powerful change agents for local and global unity.

Getting Started

Step 1: Ask all members to get a copy of this book.

Step 2: Download *The Fairness Campaign* and *The 10 Fairness Principles* from www.thebond.net.

Step 3: Post a notice and sign-up sheet at churches of New Thought (Unity, Science of Mind, or Centers for Spiritual Living), yoga or meditation groups, Whole Foods stores, and other similar types of organizations indicating that you would like to start a group to unite your community and other communities around the world.

Explain that the purpose of the group is to discuss ways to improve the local and global community and to learn to live in a more cooperative manner.

Ask people to contact you via email. If you are doing the organizing, don't hand out your home address until you know each other well.

You may also place a message through the Bond website (www .thebond.net) or Intention Experiment community (www.theinten tionexperiment.com), indicating that you would like to start a group in your area.

Step 4: Keep numbers manageable. In *The Tipping Point*, Malcolm Gladwell offers evidence that human beings were designed to work optimally in groups of 150 or fewer. If your group gets larger than that, form subgroups.

Step 5: Select a public place for your initial meetings until you really get to know the other members. Some possibilities are a cheap hall (everyone can contribute to the rental cost), a local coffee shop, a community or church hall, or even an inexpensive but quiet restaurant.

Step 6: Plan to meet weekly, if family commitments permit. A regular schedule encourages people to mark the date on their calendars and attend regularly.

Step 7: Create a structured agenda according to the weekly study guides so that people know what to expect and can come ready to contribute. Consult the weekly study guides for particular chapters to prepare for discussion topics and exercises.

Step 8: Communicate with other groups around the world. You will be able to communicate with other Bond pods via the Bond website (www.thebond.net). You'll be amazed at how quickly you'll be able to develop a worldwide virtual community of people who share your ideals and who are also intent on living in wholeness.

Group Goals

- *Begin with a model of community unity*—a society in which all members are working toward collective and unified good. Once you truly understand that you are all one, the decisions you have to make must always be for the good of all, and not simply for the people in charge, or those you like, or even those who think the way that you do.

- *Try to frame every decision in terms of its impact on your community and environment as a whole.* Will a new community project benefit everyone in the community? Does your work enhance or detract from your community? Are you educating your children to give back or just to take?

- *Create a list of each your group's resources and needs.* What talents, supplies, or general resources does each of you have available to the community? What specific needs does your community have? Can you see which talents and resources could prove most useful? Which abilities or resources can be bartered for work done by other members so that no money changes hands?

- *Invite different groups*—doctors, members of your local police force, educators—to visit your group. Explore with them ideas for reinforcing the Bond.

- *Apportion a certain number of the group's hours per week toward working on improving your community.* Volunteer to work in your local school or visit companies based in your community, exploring these ideas and how organizations can adopt them.

- *Study yourself and your true needs and invite your group members to do the same.* How much do you really need? How many new gadgets? How many new cars? What else can you do with your money?

- *Vow as a group to avoid individual ostentation.* Take a leaf from the success of Roseto, Pennsylvania, which had one of the lowest heart attack rates in America. Such was the sense of solidarity that ostentation was strictly discouraged and jealousy consequently minimized. Although rich and poor lived together, side by side, the rich did not flaunt it. Roseto was flushed with a clear sense of common purpose.

- *Vow to avoid competition with people in your community* (unless on the sports field or at the bowling alley). Does it really matter if someone makes more money than you do? Chances are, you both still face similar challenges. Also refuse to engage in *Schadenfreude*—taking delight in someone else's misfortune—and replace it with the Buddhist idea of *mudita,* or happiness in someone's good fortune.

- *Adopt the principles of the Fairness Campaign and the Ten Fairness Principles* (page 242), which aim to bring greater fairness into your community and into structures within America. The fairer you are with everyone else, the more likely they are to be fair with you.
- *Initiate new people into your group with a dialogue* (page 236). The Search for Common Ground project that united Macedonian Slavs and Albanians recommends holding periodic day-long dialogues for new members, with seasoned participants joining in at the end, or holding evening living-room dialogues that mix groups of new and old participants.

STUDY GUIDE FOR READING GROUPS AND BOND PODS

Week 1: The Superorganism: Changing Our Story

Reading Preparation: Prologue, introduction, and chapters 1 through 4 of *The Bond*

This weekly topic introduces the basic principles of the Bond: that nature's most basic drive is for global cooperation and unity, not competition. *The Bond* discloses that things in the universe are not "individual" in any sense of the term. In every area of our lives, from the smallest particles of our being to our relationships and communities, there is a Bond—a connection so integral and profound that there is no longer a clear demarcation between the end of one thing and the beginning of another.

Nature has designed us as a vast interconnected superorganism and hardwired us to share, care, and be fair, and not to compete. Everything, from subatomic particles to single-cell organisms to the most distant stars in the galaxy, all are part of this indivisible Bond. This week we will explore the basic principles of *The Bond*:

- How all of life has been designed for connection, not competition
- Why it's not the "thing," but the space *between,* that matters

Weekly Goal: To *change our story:* to understand how all of life has been designed for connection, not competition, and to explore how a new story of who we are changes how we should relate to others, in our relationships and communities.

DISCUSSION TOPICS

- Discuss the implications of new discoveries in physics showing that there is no such thing as an individual, but only relationships.

Challenge: Discuss how it would change your view of the world to think of everything in the world as one interconnected whole.

- Explore the idea that we are created "from the outside in": that our bodies are created by so many complex interactions with the environment that they cannot be considered to exist independently.

Challenge: How has your Bond with your environment changed aspects of your body and your health, compared with those of your parents?

- Discuss the idea that we are part of an "intergalactic superorganism" and that our health, our mental stability, and possibly even much of what we consider our uniquely individual behavior or motivation are partly at the whim of solar activity.

Challenge: How does this change our views of human motivation and current notions of individualism?

- We now realize that we understand the actions of others by simulating the entire experience in our brains via mirror neurons, *as though it were happening to us.*

Challenge: How does this knowledge affect our understanding of our thoughts as a wholly individual process?

- How have ideas about individualism and competition in nature affected our societal structures? How have Darwinian theories permeated our everyday lives? In what areas of your life do you feel competition most acutely?

  Challenge: How would you change societal structures to be more cooperative?

Week 2: Born to Belong

Reading Preparation: Chapters 5, 6, and 7 of *The Bond*

Humanity is profoundly tribal; we feel most at home in small clusters in which we are a part of the whole. The need to move beyond the boundaries of ourselves as individuals and to bond with a group is so primordial and necessary to a human being that it remains the key determinant of whether we remain healthy or get ill, even whether we live or die.

> **Weekly Goal:** To examine why sharing and caring is essential to our good health and explore new ways that we can flourish individually only when we move beyond speaking, seeing, and relating for just ourselves.

DISCUSSION TOPICS

- Why is the lone-wolf, all-American hero a perfect candidate for a heart attack?

  Challenge: Not all heroes in American culture represent the striving for an individualistic goal. A number demonstrate the power of emphasizing the fundamental connection between all people. One example is George Bailey in *It's a Wonderful Life,* played by James Stewart. Can you think of others?

- Why is belonging so important to people and "excessive individuation" so dangerous?

  Challenge: Remember a time when you felt left out or excluded, or a member of your group was excluded. Why did it happen? How did it affect you, emotionally and physically? What could you have done differently to promote greater unity within the group?

- How do the bad and good moods of others affect us at home or at work?

  Challenge: This week observe what happens to your mood and body language when you are in the presence of someone in a bad mood. Write it down and discuss it with the group next week.

- What are some examples of how community is the best medicine that we have, even during hard times?

  Challenge: What new groups can you join this month, or strengthen by becoming more involved with groups to which you already belong?

- We are told that human beings are innately selfish. What new evidence shows the reverse: that we are hardwired to share, care, and be fair?

  Challenge: Do something kind or altruistic for someone this week. Monitor what is happening to you, physically and emotionally, when you do.

Group Activity: Decide on three measures each one of you can take to strengthen the closeness of this group.

Week 3: The Unifying Effect of Fairness

Reading Preparation: Chapter 8 of *The Bond,* plus downloads of *The Fairness Campaign* and *The 10 Fairness Principles* from www.thebond.net and on p. 242.

Fairness burns deep within the hearts of each and every one of us; neuroscientists have even discovered an "It's not fair" spot in the human brain. Consequently ideas about fairness appear to be universal. People in most societies and citizens of all political persuasions have ideas about fairness that are surprisingly identical. By "fair," most of us mean receiving a just reward for our effort and offering an equal chance for everyone. *In our hearts we know what is fair.*

Our survival depends upon our ability to create a situation of "just desert," that we are commensurately rewarded for our efforts (or punished for our wrongdoing) and that each of us is given a turn. The extent to which any local or national community begins to fray relates to a deterioration of a sense of fairness and basic reciprocity.

As noted in chapter 8, fairness is a practice that promotes unity in any community and can be easily restored. Scientific studies show that in any society, if a culture of turn-taking falls apart with too many taking too much, all it requires is a *small group of individuals committed to strong reciprocity* to reestablish fairness and create a unified and highly cohesive community.

> **Weekly Goal:** To explore how to reestablish fairness in your life, your community, your workplace, and your country.

DISCUSSION TOPICS

- What is fairness, and how is it different from across-the-board sameness or socialism?

  Challenge: Based on your definition of fairness, identify which current local and national practices are fair and which are unfair. What would you like to change?

- Why is fairness so essential in every community, and why does it unite people?

  Challenge: Can you see any association between some of the challenges and problems in your community and fairness or unfairness?

- What happens to the social fabric in nations where there is a large sense of unfairness?

  Challenge: How have certain unfair practices in your country hurt wealthy and poor alike?

- What do you consider unfair in your own life?

  Challenge: What are some ways that you can become a game-changer to encourage fairness and fair practices in your workplace or community?

  Weekly Practice: Learn the Ten Principles of Fairness and come up with a list of ways to apply them at home, in your neighborhood, and on the job as a means of strengthening community.

Week 4: From "Me" to "We": A Slight Shift of Perception Transforms Relationships and Promotes Unity

Reading Preparation: Chapters 9 and 10 of *The Bond*

We like people who are just like us—who share our values, our attitudes, our personalities, and even our emotional dispositions—and we tend to conflict most with people who are not like us. This tendency to cluster with people who are most like us only serves to divide us from others by reinforcing our individuality, our sense that our way is the best.

Once we view ourselves as part of a larger whole, we begin to act differently toward each other. When we can learn to change our perspective and offer ourselves as a vehicle of service to the connection, we can easily find the deeper Bond always present and embrace difference within that larger definition of connection.

This week you'll learn some relating skills that will enable you to offer yourself as a vehicle of service to the pure experience of connection without judgment or prejudice shaped by your thoughts.

  Weekly Goal: To explore techniques of relating that allow you to make deep connections with anyone, even those who disagree with everything you stand for. Also, to practice the power of deep truth and candid disclosure in order to promote closeness and unity. During this type of deep sharing, the pull of wholeness builds trust and loosens attachment to entrenched positions.

DISCUSSION TOPICS

- How does our Western way of promoting the individual block us from seeing another version of reality? What can survivors of the 2004 tsunami teach us about taking a more holistic view of our actions?

  Challenge: Think of some of your actions and how they impact your community.

- Discuss the art of *Sawubona* and what it means to say to someone else, "We see you."

  Challenge: Divide into pairs and say "We see you" and reply with "We see you too." At that moment, vow to do whatever you can to allow your partner to thrive. How does that change your view of the relationship?

- What is "aerial vision"?

  Challenge: Think of a time when you and another person vehemently disagreed on something. What was his or her version of reality and your version, and where is the truth of both sides?

- Discuss the practice of dialogue, how it differs from ordinary discussion and how it overcame polarization among the pro-life and pro-choice activists in Cambridge, Massachusetts.

  Challenge: Discuss some of the assumptions you make about your version of what's "real." How much of it is based on cultural conditioning and belief?

- Why is deep sharing so powerful in relationships? How does it help us forgive and reestablish connection, as it did with a former member of the Hitler youth movement and the daughter of a Holocaust survivor, as mentioned in chapter 10?

  Challenge: Divide into pairs and practice sharing deeply about things you really care about. What do you notice about how you feel about the other person?

  Weekly Practice: In your group, practice having a dialogue about a contentious subject (abortion, the Tea Party, gun control, etc.). Remember the rules:

- Don't reach a decision or have a debate.
- Take turns speaking.
- Be alert to your own reactions when something is said that you don't agree with.
- Be fully present.
- Don't judge.

Week 5: People Who Fire Together Wire Together

Reading Preparation: Chapter 11 of *The Bond*

This week we'll use the Robber's Cave experiment to examine the power of a larger, communal (superordinate) goal and how it can transform the atmosphere in your neighborhood or office from "me against them" to "all of us pulling together." It may also touch upon the power of the larger goal to revitalize any social situation, in your office or neighborhood, and create a close-knit and cooperative community. Shared activity also creates a ramped-up endorphin release, raising our pain threshold, improving our individual efforts, and ultimately raising our game. You may also wish to explore specific practices to turn your community into a communal savings bank to ease individual worries during hard times.

DISCUSSION TOPICS

- What are the lessons of the Robber's Cave experiment for today?

  Challenge: Think of a superordinate goal that your group could pursue in your community in order to unite people who are presently polarized.

- Why does a superordinate goal unite people so well?

  Challenge: What practices might you suggest to transform the atmosphere in your office from "me against them" to "all of us pulling together"?

- What tips can you gain from South Africa's rugby team and Oxford's rowing team, the Chilean miners, the Tailholt experience with their community water pipeline, and Nour Hakki, the Syrian translator, about creating a more united community?

Challenge: Can you think of three communal practices to establish in your neighborhood as a way of making your neighborhood safer and more vibrant, and your neighbors feel closer?

- In Okinawa the Japanese like to handle their money through a *moai,* turning a trusted set of friends or neighbors into a savings bank. How could we adapt this idea in our neighborhoods?

Challenge: Think of three ways to create a neighborhood savings bank to help each other when times are hard. The bank doesn't have to hold money; neighbors can help each other by raking leaves, repairing a fence, or even bartering, trading work or products instead of money.

Weekly Practice: As a group, decide on a specific project in your community about which people of all religious beliefs, cultural backgrounds, and political persuasions feel equally passionate. Set up a committee to begin working on getting the project under way. Once you have the basics of the project, organize another committee whose job is to invite people of differing beliefs or cultures to join. Take note of whether working together for a common goal brings you closer together.

Week 6: Generosity and Fairness: The Global Game-Changer

Reading Preparation: Chapter 12 of *The Bond*

This week explores ways that each of you within your group or congregation can become "spiritual citizens" and transform your own life's purpose from "me and mine alone" to "ours." The session should examine how each of us can become a powerful change agent through tiny "do something" actions and investigate the extraordinary and infectious power of generosity to transform a business or neighborhood. Simple everyday acts of generosity enable you to become a powerhouse of change, permanently shifting the culture of greed and materialism around you.

Weekly Goal: To explore the extraordinary and infectious power of generosity to create inherent trust in your community or workplace.

DISCUSSION TOPICS

- What are the most important lessons to be learned from the experiences of Nipun Mehta, who transformed from a classic Silicon Valley turk on a six-figure salary to a global game-changer running the international CharityFocus?

  Challenge: Which are the most important ways you could become a change agent in your daily life?

- How can each one of us set off a contagion of giving and cooperative behavior in our neighborhood or office?

  Challenge: Think of several ways that you can use generosity as an extraordinary and infectious power to create inherent trust in your neighborhood.

- Describe why one tiny act of kindness—leaving a bit of change in a Coke machine—caused a wave of generosity throughout a vast corporation and affected an entire community. What happens to a culture of selfishness in the face of a few change agents who reestablish generosity and reciprocity?

  Challenge: Design several acts you can do that might set off a contagion of change.

- How can partnerships (like the John Lewis Partnership) be used to unite communities?

  Challenge: Can you reorganize a project you are currently working on as a partnership with no one in charge?

Weekly Practice: Design three specific pay-it-forward actions in your neighborhood or workplace that will set off a contagion of giving and cooperative behavior. Report back to the group on the results.

Week 7: From Local to Global

Reading Preparation: Chapters 8, 11, and 12 of *The Bond*

In our final week we'll examine ways to move from the old zero-sum paradigm ("I win, you lose") to the six principles of the Bond ("I win, *and* we all win") globally. We'll also explore the power of designing superordinate goals to connect with and evolve with other groups around the world. This lesson also focuses on the Nash Equilibrium (what's best for me *and* the group) to design goals that promote unity with others from different cultures or belief systems.

This can be illustrated by the story of the mathematician John Nash, as portrayed in the movie *A Beautiful Mind* (detailed in chapter 8 of *The Bond*), when Nash realized that Adam Smith's model of "every man for himself" is wrong. The best result comes, he says, when everyone in the group does what's best for himself *and* the group.

> **Weekly Goal:** To explore how moving from a paradigm of exclusion ("I win, you lose," Us versus Them) to inclusion ("I win *and* we all win,"Us *plus* Them) offers a powerful tool for connection and to bring these ideas to the wider community.

THE "I WIN, YOU LOSE" PARADIGM

- Life is a zero-sum game (I win only if you lose).
- What's in it for me?
- I must win, dominate, or be first in order to feel good about myself.
- I take care of my own. Otherwise, it's every man for himself.
- It's Us versus Them.
- I do best for society by looking after Number One.

THE BOND PARADIGM

- I win *when and only when* you and I win.
- I feel best when I connect, whatever it takes.
- How can I serve you?

- We're all in this together, and we will pull together through thick and thin.
- It's Us *plus* Them.
- We do best for society by looking out for ourselves *and* the rest of our community.

DISCUSSION TOPICS

- Discuss the bar scene in the film *A Beautiful Mind,* detailed in chapter 9 of *The Bond,* when Nash realizes that the best action you can take is the one that's not only what's best for yourself but is also best for the group.

  Challenge: What are some ways to apply this idea to your workplace or community?

- What do the lessons of the Nash Equilibrium and other examples of game theory teach us about our current paradigm ("I win, you lose")?

  Challenge: Discuss one area in your life and community that is currently "every man for himself," and redesign it as "every man for himself *and* the group." How would it work then?

- Engage in a bit of blue-skying. What would it look like to have businesses that work together and not in competition?

  Challenge: Discuss ways to incorporate these ideas into your own workplace or a community organization.

- Discuss the old and new paradigms listed above and where they are present in our neighborhoods and communities, businesses and political system.

  Challenge: Can you design new ideas in your communities that incorporate the "Us *plus* Them" ideas of the Bond paradigm into our everyday lives?

- How might this new paradigm be used politically to improve aspects of society?

Challenge: Describe how the "Us *plus* Them" model might unite people of different political persuasions.

Weekly Practice: Design a superordinate "Us plus Them" goal to encompass the needs of your community and then several communities outside of your own. Begin making plans to put it into practice. Continue reporting back to the group on your progress.

ACKNOWLEDGMENTS

A book is always a Bond—a communion of the thoughts and work of many individuals.

This book was assembled from a large library of scientific literature, popular books, and journalism, plus interviews with most of the scientists and other parties named in this book. I am grateful to Daniel Batson, Don Beck, Orland Bishop, Lindsay Browning, Nicholas Christakis, Andrew Colman, Germaine Cornélissen, Ernst Fehr, Graham Fleming, Nour Hakki, Franz Halberg, Randy Jirtle, Yoko Masashiro, Nipun Mehta, James O'Dea, Samuel Oliner, Hal Puthoff, Giacomo Rizzolatti, Trishna Shah, Leonard Syme, Vlatko Vedral, and Robert Waterland for agreeing to be interviewed and, in most instances, for reviewing and correcting any errors in those sections of the manuscript that describe their work. I'm also grateful to Fukiko Kai for generously setting up a three-way telephone link with Yoko Masashiro and translating our interview.

I'd also like to thank my eldest daughter, Caitlin, who holds a degree in psychology. She first recommended that I examine the contact hypothesis.

I owe a special debt of gratitude to my entire team at Free Press—Dominick Anfuso, Suzanne Donahue, Carisa Hays, Meg Cassidy, and Donna Loffredo both for their incredible support of me and my work and for their openness to new ideas. I especially appreciated the work of the eagle-eyed Judith Hoover who improved this manuscript in countless ways.

Sharyn Wong, my own company's production editor, performed some eleventh-hour help in assembling my bibliography and notes; Pavel Mikoloski again set up an extensive publicity tour for the hardcover edition; and Kyle Gibson provided administrative help on the book and my tour. Andrew Coleman, my UK copyeditor, performed an invaluable service by helping to fact-check this version.

I am also grateful to Danny and Heather Benor, my foreign agents, for all their diligent effort in insuring that my books reach audiences around the world. As always, I owe an enormous debt of gratitude to my agent, Russell Galen, the constant gardener of my work and best interests for more than twenty years.

Special mention should be made of my sister-in-law, Trina Hubbard, for buying me a slow cooker, without which this book could not have been written. And a tribute paid to our little dog, Ollie, the inspiration for a number of passages in this book, who died between the publication of the hardcover and paperback editions of this book.

I am especially blessed to have such an extraordinary editor in Leslie Meredith, whose painstaking editing and dedication to the improvement of this book was nothing short of astonishing. I am particularly grateful for the enormous faith she has exhibited in me on more occasions than I can detail, and for her talent for teasing out of me the full story of what I wanted to tell.

I must especially thank both my daughters, Caitlin and Anya, who teach me every day to move past superficial difference to the place of pure connection.

Finally, this book owes its greatest debt to my husband, Bryan Hubbard, who once again has nurtured me in every way through every stage of this book, whose brilliant insights always encouraged me to go deeper, and through whom, on a daily basis, I experience the power of the space in between.

NOTES

PROLOGUE

1. Certain details of this incident have been changed.

INTRODUCTION

1. A. Smith, An Inquiry into the Nature and Cause of the Wealth of Nations (W. Strahan and T. Cadell, 1776).
2. T. R. Malthus, *An Essay on the Principle of Population As It Affects the Future Improvement of Society* (1798; reprint, Oxford World's Classics), 13.
3. C. Darwin, *On the Origin of Species by Means of Natural Selection, or the Preservation of the Favoured Races in the Struggle for Life* (John Murray, 1859).
4. A. R. Wallace, "Mr. Darwin's Metaphors Liable to Misconception," 1868, cited in D. Todes, "Global Darwin: Contempt for Competition," *Nature* 462, no. 5 (2009): 36–37. New readings of Darwin suggest that his central metaphor was "accommodation" rather than survival struggle; he came to believe that life endlessly transformed and adapted in response to the environment.
5. M. Elshakry, "Global Darwin: Eastern Enchantment," *Nature* 461, no. 2 (2009): 1200–1201.
6. J. Pusey, "Global Darwin: Revolutionary Road," *Nature* 462, no. 12 (2009): 162–63; J. Buchenau, "Global Darwin: Multicultural Mergers," *Nature* 462, no. 19 (2009): 284–85.
7. See R. Dawkins, *The Selfish Gene* (Oxford University Press, 1989).
8. Ibid., 2. See also Rupert Sheldrake, *The Presence of the Past* (HarperCollins, 1988), 83–85.
9. B. Lipton, *Spontaneous Evolution: Our Positive Future and a Way to Get There from Here* (Hay House, 2009).
10. P.C. Gøtzsche et al. "Ghost authorship in Industry-initiated Randomised Trials," PLoS Med, 2007; 4 (1): e19.
11. D. Sewell, *The Political Gene: How Darwin's Ideas Changed Politics* (Picador, 2009).
12. L. McTaggart, *The Field: the Quest for the Secret Force of the Universe* (HarperCollins, 2002).

NOTES

CHAPTER 1

1. G. Engel et al., "Evidence for Wavelike Energy Transfer through Quantum Coherence in Photosynthetic System," letter, *Nature* 446, no. 112 (2007): 782–86.
2. S. Weinberg, "What Is an Elementary Particle?," *Beam Line: A Periodical of Particle Physics* 27 (Spring 1997): 17–21.
3. W. Heisenberg, "Talk to the German Physical Society," 1975, cited in S. Weinberg, "What Is an Elementary Particle?," emphasis added.
4. S. Weinberg, "What Is an Elementary Particle?"
5. W. Heisenberg, "Über den Anschaulichen Inhalt der Quantentheoretischen Kinematik und Mechanik," *Zeitschrift für Physik.* 43 (1927): 172–98.
6. John D. Barrow, *The Book of Nothing* (Jonathan Cape, 2000), 216.
7. H. E. Puthoff, "Ground State of Hydrogen as a Zero-point-fluctuation-determined State," *Physical Review D* 35 (1987): 3266.
8. B. Haisch, A. Rueda, and H. E. Puthoff, "Inertia as a Zero-point-field Lorentz Force," *Physical Review A* 49, no. 2 (1994): 678–94; B. Haisch, A. Rueda, and H. E. Puthoff, "Physics of the Zero-point Field: Implications for Inertia, Gravitation and Mass," *Speculations in Science and Technology* 20 (1997): 99–114.
9. B. Haisch, A. Rueda, and H. E. Puthoff, "Advances in the proposed electromagnetic zero-point field theory of inertia," paper presented at AIAA 98–3143, Advances ASME/SAE/ASEE Joint Propulsion Conference and Exhibit, July 13–15 1998, Cleveland, Ohio.
10. J. A. Wheeler, "Information, Physics, Quantum: The Search for Links," in W. Zurek, ed., *Complexity, Entropy, and the Physics of Information* (Addison-Wesley, 1990).
11. J. A. Wheeler, *Mathematical Foundations of Quantum Theory*, A.R. Marlow, ed. (Academic Press, 1978.)
12. Nonlocality was considered to be proven by Alain Aspect et al.'s experiments in Paris in 1982: A. Aspect et al., "Experimental Realization of Einstein-Podolsky-Rosen-Bohm Gedankenexperiment: A New Violation of Bell's Inequalities," *Physical Review Letters,* 49 no. 2 (1982): 91–4. V. Jacques et al., "Experimental Realization of Wheeler's Delayed-Choice Gedankenexperiment," *Science* 315, no. 5814 (2007): 966–68.
13. J. A. Wheeler, speaking on "The Anthropic Universe," on ABC Radio National, *The Science Show*, February 18, 2006.

CHAPTER 2

1. L.H. Lumey, "Decreased Birthweights in Infants After Maternal in Utero Exposure to the Dutch Famine of 1944–1945," *Paediatric and Perinatal Epidemiology* 6, no. 2 (1992): 240–53; G. Kaati et al., "Cardiovascular and Diabetes Mortality Determined by Nutrition During Parents' and Grandparents' Slow Growth Period," *European Journal of Human Genetics* 10 (2002): 682–88. See Harald Gaier, "Beyond the Selfish Gene," *What Doctors Don't Tell You,* August 2008, 22.
2. G. Wolff et al., "Maternal Epigenetics and Methyl Supplements Affect Agouti Gene Expression in Avy/a Mice," *FASEB Journal* 12 (August 1998): 949–57.

NOTES

3. R. A. Waterland and R.L. Jirtle, "Transposable Elements: Targets for Early Nutritional Effects on Epigenetic Gene Regulation,"*Molecular Cell Biology* 23 (2003): 5293–300.

4. Quoted in L. A. Pray, "Epigenetics: Genome, Meet Your Environment," *The Scientist* 18, no. 13 (2004): 14–20.

5. J. Lamarck, *Philosophie Zoologique, ou Exposition des Considérations Relatives à l'Histoire Naturelle des Animaux* (Dentu, 1809).

6. A. C. Faberge and J. D. Mohler, "Breakage of Chromosomes Produced by Ultraviolet Radiation in Drosophila," *Nature* 169, no. 4294 (1952): 278–79.

7. C. H. Waddington, "Genetic Assimilation of an Acquired Character," *Evolution* (1953): 118–26; C. H. Waddington, *The Evolution of an Evolutionist* (Cornell University Press, 1975).

8. J. D.Watson and F. H. Crick, "Molecular Structure of Nucleic Acids: A Structure for Deoxyribose Nucleic Acid." *Nature* 171, no. 4356 (1953): 737–38.

9. The biologist Bruce Lipton was one of the first to make this connection in his groundbreaking book, *The Biology of Belief: Unleashing the Power of Consciousness, Matter, & Miracles* (Hay House, 2008).

10. Ibid., chapter 3, 75–94.

11. Ibid., 75–6.

12. M. Szyf, "DNA Memethylation and Cancer: Therapeutic Implications," *Cancer Letters* 10, no. 211 (2004): 133–43.

13. J. N. Ou et al., "Histone Deacetylase Inhibitor Trichostatin A Induces Global and Gene-specific DNA Demethylation in Human Cancer Cell Lines," *Biochemical Pharmacology* 73, no. 9 (2007):1297–307.

14. I. C. Weaver et al., "Epigenetic Programming by Maternal Behavior," *Nature Neuroscience* 7, no. 8 (2004): 847–54; I. C. Weaver et al., "Maternal Care Effects on the Hippocampal Transcriptome and Anxiety-mediated Behaviors in the Offspring That Are Reversible in Adulthood," *Proceedings of the National Academy of Sciences USA* 103, no. 9 (2006): 3480–85.

15. Jacques E. Rossouw et al., "Risks and Benefits of Estrogen Plus Progestin in Healthy Postmenopausal Women: Principal Results From the Women's Health Initiative Randomized Controlled Trial," *Journal of the American Medical Association,* 288 (2002): 321–33.

16. R. Gramling et al., "Hormone Replacement Therapy, Family History and Incident Invasive Breast Cancer Among Postmenopausal Women in the Women's Health Initiative," *Epidemiology* 20 (2009): 752–56.

17. P. O. McGowan et al., "Promoter-wide Hypermethylation of the Ribosomal RNA Gene Promoter in the Suicide Brain," *PLoS ONE* 3, no. 5 (2008): e-2085.

18. P. O. McGowan et al, "Epigenetic Regulation of the Glucocorticoid Receptor in Human Brain Associates with Childhood Abuse," *Nature Neuroscience,* 12 (2009): 342–48.

19. J. Mill et al., "Epigenomic Profiling Reveals DNA-Methylation Changes Associated with Major Psychosis," *American Journal of Human Genetics* 82, no. 3 (2008): 696–711.

20. A. Fischer et al., "Recovery of Learning and Memory Is Associated with Chromatin Remodelling," *Nature* 447 (May 10, 2007): 178–182.

21. J. Arai et al., "Transgenerational Rescue of a Genetic Defect in Long-Term Poten-

tiation and Memory Formation by Juvenile Enrichment," *Journal of Neuroscience* 29, no. 5 (2009): 1496–502.

22. G. Kaati et al, "Transgenerational Response to Nutrition, Early Life Circumstances and Longevity," *European Journal of Human Genetics* 15 (2007): 784–90.

23. M. E. Pembrey et al., "Sex-specific, Male-line Transgenerational Responses in Humans," *European Journal of Human Genetics* 14 (2006): 159–166.

24. A. Fraser et al., "Association of Maternal Weight Gain in Pregnancy with Offspring Obesity and Metabolic and Vascular Traits in Childhood," *Circulation* 121 (2010): 2557–64.

25. R. Alleyne, "Britain's 'Me Culture' Making Us Depressed," *The Telegraph*, November 6, 2009.

26. J. Chiao and K. Blizinsky, "Culture-Gene Coevolution of Individualism-Collectivism and the Serotonin Transporter Gene," *Proceedings of the Royal Society B: Biological Sciences* 277, no 1681 (2010): 529–37.

27. J. Cairns, "The Origin of Mutants," *Nature,* 335 (1988): 142–45.

28. D. Charlesworth et al., "Origin of Mutants Disputed," *Nature,* 336 (1988): 525; R. E. Lenski and J. E. Mittler, "The Directed Mutation Controversy and Neo-Darwinism," *Science,* 259 (1993): 188–94; plus related correspondence in *Science* 260 (1993): 1221–24, 1958–60.

29. B. Lipton, *Spontaneous Evolution,* 149–50.

30. J. A. Shapiro, "Mobile DNA and Evolution in the 21st Century," *Mobile DNA* 1 (2010): 4.

31. J. A. Shapiro. "Bacteria Are Small but Not Stupid: Cognition, Natural Genetic Engineering and Socio-bacteriology," *Studies in History and Philosophy of Science Part C: Studies in History and Philosophy of Biological and Biomedical Sciences* 38, no 4 (2007): 807–19.

CHAPTER 3

1. F. Halberg, "Historical Encounters between Geophysics and Biomedicine Leading to the Cornélissen-series and Chronoastrobiology," In: Schröder, W., ed. *Long- and Short-Term Variability in Sun's History and Global Change* (Bremen: Science Edition [2465] 2000): 272; F. Halberg et al., "Cycles Tipping the Scale between Death and Survival (= 'Life')," *Progress of Theoretical Physics,* suppl. 173 (2008): 153–81.

2. I have examined this in detail in *The Intention Experiment* (Free Press, 2007), 100–101.

3. L. E. Scheving and F. Halberg, *Chronobiology: Principles and Applications to Shifts in Schedules* (Kluwer Academic, 1981).

4. W. Hrushesky et al. "Cisplatine Chronotolerance," *Proceedings of the Conference on Combined Modalities: Chemotherapy/Radiotherapy* [567], abstract CP-V-3 (1978): 58.

5. G. Cornélissen et al., "Is a Birth-month-dependence of Human Longevity Influenced by Half-yearly Changes in Geomagnetics?" *Proceedings of the XXV Annual Seminar, "Physics of Auroral Phenomena,"* Apatity, February 26-March 1, 2002. Apatity: Polar Geophysical Institute, Kola Science Center, Russian Academy of Science; [2657b] (2002):. 161–65.

NOTES

6. V. N. Bingia and A. V. Savin, "Effects of Weak Magnetic Fields on Biological Systems: Physical Aspects," *Uspekhi Fizicheskikh Nauk* 173, no. 3 (2003): 265–300.
7. Interview with Germaine Cornélissen, February 11, 2010.
8. V. M. Petro et al., "An Influence of Changes of Magnetic Field of the Earth on the Functional State of Humans in the Conditions of Space Mission," paper presented at the International Symposium Computer Electro-Cardiograph on Boundary of Centuries, Moscow, Russian Federation, April 27–30, 1999.
9. Germaine Cornélissen et al., "Chronomes, Time Structures, for Chronobioengineering for 'a Full Life,'" *Biomedical Instrumentation and Technology* 33, no. 2 (1999): 152–87.
10. V. N. Oraevskii et al., "Medico-Biological Effect of Natural Electromagnetic Variations," *Biofizika* 43, no. 5 (1998): 844–48; V. N. Oraevskii et al., "An Influence of Geomagnetic Activity on the Functional Status of the Body," *Biofizika* 43, no. 5 (1998): 819–26.
11. F. Halberg et al., "Cross-spectrally Coherent about 10-, 5- and 21-year Biological and Physical Cycles, Magnetic Storms and Myocardial Infarctions," *Neuroendrocrinology Letters* 21 (2000): 233–58; I. Gurfinkel et al., "Assessment of the Effect of a Geomagnetic Storm on the Frequency of Appearance of Acute Cardiovascular Pathology," *Biofizika* 43, no. 4 (1998): 654–58; J. Sitar, "The Causality of Lunar Changes on Cardiovascular Mortality," *Casopis Lekaru Ceskych* 129 (1990): 1425–30.
12. M. N. Gnevyshev, "Essential Features of the 11-year Solar Cycle," *Solar Physics* 51 (1977): 175–82.
13. G. Cornélissen et al., "Non-photic Solar Associations of Heart Rate Variability and Myocardial Infarction," *Journal of Atmospheric and Solar-terrestrial Physics* 64 (2002): 707–20.
14. R. M. Baevsky et al. "Meta-analyzed Heart Rate Variability, Exposure to Geomagnetic Storms, and the Risk of Ischemic Heart Disease," *Scripta Medica (Brno)* 70, no. 4–5 (1997): 201–6; G. Cornélissen et al. "From Various Kinds of Heart Rate Variability to Chronocardiology," *American Journal of Cardiology* 66 (1990): 863–68.
15. Baevsky et al. "Meta-analyzed Heart Rate Variability"; Petro et al., "An Influence of Changes of Magnetic Field."
16. Cornélissen et al., "Chronomes, Time Structures."
17. A. R. Allahverdiyev et al., "Possible Space Weather Influence on Functional Activity of the Human Brain," paper presented at the Space Weather Workshop: Looking towards a European Space Weather Programme, Noordwijk, The Netherlands, December 17–19, 2001.
18. E. Babayev, "Some Results of Investigations on the Space Weather Influence on Functioning of Several Engineering-technical and Communication Systems and Human Health," *Astronomical and Astrophysical Transactions* 22, no. 6 (2003): 861–67; G. Y. Mizun and P. G. Mizun, "Space and Health," *Znanie* (1984).
19. Allahverdiyev et al., "Possible Space Weather Influence"; Babayev, "Some Results of Investigations"; Mizun and Mizun, "Space and Health."
20. Avi Raps et al., "Geophysical Variables and Behavior: LXIX. Solar Activity and Admission of Psychiatric Inpatients," *Perceptual and Motor Skills* 74 (1992): 449; H. Friedman et al., "Geomagnetic Parameters and Psychiatric Hospital Admis-

sions," *Nature* 200 (1963): 626–28; E. Stoupel, "Relationship Between Suicide and Myocardial Infarction with Regard to Changing Physical Environmental Conditions," *International Journal of Biometeorology* 38, no. 4 (1994): 199–203; E. Stoupel et al., "Clinical Cosmobiology: The Lithuanian Study, 1990–1992," *International Journal of Biometerology* 38 (1995): 204–8; E. Stoupel et al., "Suicide-Homicide Temporal Interrelationship, Links with Other Fatalities and Environmental Physical Activity," *Crisis* 26 (2005): 85–89.

21. R. Becker, *The Body Electric: Electromagnetism and the Foundation of Life* (Quill, 1985).

22. F. Halberg, "Chronomics of Autism and Suicide," *Biomedicine and Pharmacotherapy* 59 (2005): S100–S108.

23. M. Mikulecky, "Lunisolar Tidal Waves, Geomagnetic Activity and Epilepsy in the Light of Multivariate Coherence," *Brazilian Journal of Medicine* 29, no. 8 (1996): 1069–72; E. A. McGugan, "Sudden Unexpected Deaths in Epileptics: A Literature Review," *Scottish Medical Journal* 44, no. 5 (1999): 137–39; Y. Bureau and M. Persinger, "Decreased Latencies for Limbic Seizures Induced in Rats by Lithium-Pilocarpine Occur When Daily Average Geomagnetic Activity Exceeds 20 Nanotesla," *Neuroscience Letters* 192 (1995): 142–44; A. Michon and M. Persinger, "Experimental Simulation of the Effects of Increased Geomagnetic Activity upon Nocturnal Seizures in Epileptic Rats," *Neuroscience Letters* 224 (1997): 53–56.

24. M. Persinger, "Sudden Unexpected Death in Epileptics Following Sudden, Intense Increases in Geomagnetic Activity: Prevalence of Effect and Potential Mechanisms," *International Journal of Biometeorology* 38 (1995): 180–87; R. P. O'Connor and M. A. Persinger, "Geophysical Variables and Behavior LXXXII: A Strong Association between Sudden Infant Death Syndrome (SIDS) and Increments of Global Geomagnetic Activity. Possible Support for the Melatonin Hypothesis," *Perceptual and Motor Skills* 84 (1997): 395–402.

25. T. Long et al., "Relationship of Daily Geomagnetic Activity to the Occurrence of Temporal Lobe Seizures in an Epilepsy Monitoring Unit, paper presented at American Epilepsy Society, May 1996. Abstract published in *Epilepsia* 36(S4) (1996): 94.

26. Cornélissen et al., "Chromosomes, Time Structures."

27. M. Berk et al., "Do Ambient Electromagnetic Fields Affect Behavior? A Demonstration of the Relationship between Geomagnetic Storm Activity and Suicide," *Bioelectromagnetics* 27 (2006): 151–55.

28. Interview with G. Cornélissen, February 11, 2010.

29. Interview with G. Cornélissen, February 11, 2010.

30. S. Starbuck et al., "Is Motivation Influenced by Geomagnetic Activity?," *Biomedicine and Pharmacotherapy* 56 (2002): 289s—297s.

31. A. Krivelyova and C. Robotti, "Playing the Field: Geomagnetic Storms and the Stock Market," Federal Reserve Bank of Atlanta Working Paper 2003–5b, October 2003.

32. TSAA Newsletter, April 1996.

33. F. Halberg et al., "Subtraction and Addition of Heliomagnetics and Human Violence: Time Course of Transyears in Terrorism Mimics Their Cosmos," paper presented at the first International Workshop, Physiology of Adaptation and Quality of Life: Problems of Traditional Medicine and Innovation, People's Friendship University of Russia, Moscow, May 14–16, 2008.

34. A. Lieber, "Human Aggression and the Lunar Synodic Cycle," *Journal of Clinical Psychiatry* 39 no. 5 (1978): 385–92; A. Lieber, and C. R. Sherin, "Homicides and the Lunar Cycle: Toward a Theory of Lunar Influence on Human Behavior," *American Journal of Psychiatry* 129 (1972): 69–74.

35. C. Bhattacharjee, "Do Animals Bite More During a Full Moon? Retrospective Observational Analysis," *British Medical Journal* 321 (2000): 1559–61.

36. D. Templer and D. Veleber, "The Moon and Madness: A Comprehensive Perspective," *Journal of Clinical Psychology,* 36, no 4 (1980): 865–68; C. E. Climent and R. Plutchik, "Lunar Madness: An Empirical Study," *Comprehensive Psychiatry* 18, no. 4: 369–74.

37. N. Kollerstrom and B. Steffert, "Sex Difference in Response to Stress by Lunar Month: A Pilot Study of Four Years' Crisis-call Frequency," *BMC Psychiatry* 3 (2003): 20.

38. R. D. Neal and M. Colledge, "The Effect of the Full Moon on General Practice Consultation Rates," *Family Practice* 17, no 6 (2000): 472–4.

39. C. T. Russell et al., "The Permanent and Induced Magnetic Dipole Moment of the Moon," *Proceedings of the Fifth Lunar Conference* 3 (1974): 2747–60.

40. A.M. Forte and J.X. Mitrovica, "A Resonance in the Earth's Obliquity and Precession Over the Past 20Myr Driven by Mantle Convection," *Nature* 390 (1997): 676–80.

41. J. Mitrovica, quoted in a University of Toronto press release, "Other Planets Influence Earth's Climate, U of T Scientist Says," December 17, 1997.

42. M. Mikulecky et al., "Lunisolar Tidal Waves, Geomagnetic Activity and Epilepsy in the Light of Multivariate Coherence," *Brazilian Journal of Medical and Biological Research* 8 (1996): 1069.

43. C. D. Murray and S. F. Dermott, *Solar System Dynamics* (Cambridge University Press, 1999).

44. J. Mitrovica, quoted in "Other Planets Influence Earth's Climate."

45. Halberg, "Cycles Tipping the Scale between Death and Survival."

CHAPTER 4

1. G. Di Pellegrino et al, "Understanding Motor Events: A Neurophysiological Study," *Experimental Brain Research* 91 (1992): 176–80.

2. G. Rizzolatti and L. Craighero, "Mirror Neuron: A Neurological Approach to Empathy," in *Neurobiology of Human Values,* ed. J. P. Changeux et al. (Springer-Verlag, 2005), 108–23.

3. L. Fogassi et al, "Parietal Lobe: From Action Organization to Intention Understanding," *Science* 308 (2005): 662–67.

4. Ibid.

5. S. V. Shepherd et al., "Latency of Social-cued Attention Signals in Macaque Area LIP," paper presented at Neuroscience 2007, the 37th annual meeting of the Society for Neuroscience, San Diego, 2007; C. Keysers and L. Fadiga, "The Mirror Neuron System: New Frontiers," *Social Neuroscience* 3, nos. 3–4 (2008): 193–98.

6. Keysers and Fadiga, "The Mirror Neuron System"; C. Keysers et al., "Audiovisual Mirror Neurons and Action Recognition," *Experimental Brain Research* 153 (2003): 628–36.

7. G. Rizzolatti and L. Craighero, "The Mirror Neuron System," *Annual Review of Neuroscience* 27 (2004): 169–92.

8. A. Meltzoff and W. Prinz, *The Imitative Mind: Development, Evolution, and Brain Bases* (Cambridge University Press, 2002).

9. Di Pellegrino et al., "Understanding Motor Events."

10. V. Gallese et al., "Action Recognition in the Premotor Cortex," *Brain* 119, no. 2 (1996): 593–609.

11. V. Gallese, "The 'Shared Manifold' Hypothesis: From Mirror Neurons to Empathy," *Journal of Consciousness Studies* 8 (2001): 33–50.

12. C. van der Gaag, "Facial Expressions: What the Mirror Neuron System Can and Cannot Tell Us," *Social Neuroscience* 2, nos. 3–4 (2007): 179–222; S. Pichon et al., "Emotional Modulation of Visual and Motor Areas by Dynamic Body Expressions of Anger," *Social Neuroscience* 3, nos. 3–4 (2008): 199–212.

13. M. Iacobani et al., "Cortical Mechanisms of Human Imitation," *Science* 286 (December 24, 1999): 2526–28.

14. B. Wicker et al., "Both of Us Disgusted in *My* Insula: The Common Neural Basis of Seeing and Feeling Disgust," *Neuron* 40 (2003): 655–64.

15. V. Gazzola et al., "The Anthropomorphic Brain: The Mirror Neuron System Responds to Human and Robotic Actions," *NeuroImage* 35 (2007): 1674–84.

16. Keysers and Fadiga, "The Mirror Neuron System."

17. C. Keysers and V. Gazzola, "Towards a Unifying Neural Theory of Social Cognition," in *Progress in Brain Research*, vol. 156, ed. S. Anders et al. (Elsevier, 2006), 379–401.

18. C. Keysers et al., "A Touching Sight: SII/VV Activation During the Observation and Experience of Touch," *Neuron* 42 (2004): 335–46.

19. V. Gazzola et al., "Aplastics Born Without Hands Mirror the Goal of Hand Actions with Their Feet," *Current Biology* 17 (2007): 1235–40.

20. C. Catmur et al., "Sensorimotor Learning Configures the Human Mirror System," *Current Biology* 17 (2007): 1527–31.

21. Keysers and Gazzola, "Towards a Unifying Neural Theory."

22. Keysers, "A Touching Sight."

23. Ibid.

24. Interview with G. Rizzolatti, February 8, 2010.

25. W. D. Hutchison et al., "Pain-related Neurons in the Human Cingulate Cortex," *Nature Neuroscience* 2 (1999): 403–5.

26. S. Han et al., "Empathic Neural Responses to Others' Pain Are Modulated by Emotional Contexts," *Human Brain Mapping* 30 (2009), 3227–37.

27. T. Singer et al., "Empathy for Pain Involves the Affective but Not Sensory Components of Pain," *Science* 303, no. 5661 (2004): 1157–62.

28. Interview with G. Rizzolatti, February 8, 2010.

29. M. Jabbi et al., "Empathy for Positive and Negative Emotions in the Gustatory Cortex," *NeuroImage* 34 (2007): 1744–53; V. Gazzola et al., "Empathy and the Somatotopic Auditory Mirror System in Humans," *Current Biology* 16 (2006): 1824–29.

30. S. D. Preston et al., "The Neural Substrates of Cognitive Empathy," *Social Neuroscience* 2, No. 3–4 (2007): 254–75.

31. A. Schore, *Affect Regulation and the Origin of Self* (Psychology Press, 1999).

NOTES

32. G. Schwartz, "Energy Medicine and Bioenergy Feedback," presentation at FutureHealth Conference, Palm Springs, February 5–9, 1998.

33. J. Chiltern Pierce, *Magical Child* (Bantam Books, 1980)

34. L. McTaggart, "The Love Study," in *The Intention Experiment* (Free Press, 2007).

35. For details of all the studies, see McTaggart, *The Intention Experiment.*

36. L. Standish et al., "Electroencephalographic Evidence of Correlated Event-Related Signals between the Brains of Spatially and Sensory Isolated Human Subjects," *Journal of Alternative and Complementary Medicine* 10, no. 2 (2004): 307–14.

37. M. Kittenis et al., "Distant Psychophysiological Interaction Effects between Related and Unrelated Participants," Proceedings of the Parapsychological Association Convention, 2004, 67–76, as reported in D. Radin, *Entangled Minds* (Paraview, 2006): 138–39.

38. This study is recounted in detail in McTaggart, *The Intention Experiment,* chapter 4.

39. D. I. Radin and M. J. Schlitz, "Gut Feelings, Intuition and Emotions: An Exploratory Study," *Journal of Alternative and Complementary Medicine* 11, no. 5 (2005): 85–91.

40. D. I. Radin, "Event-related EEG Correlations between Isolated Human Subjects," *Journal of Alternative and Complementary Medicine* 10 (2004): 315–24; M. Cade and N. Coxhead, *The Awakened Mind,* 2nd ed. (Element, 1986); S. Fahrion et al., "EEG Amplitude, Brain Mapping and Synchrony in and between a Bioenergy Practitioner and Client During Healing," *Subtle Energies and Energy Medicine* 3, no. 1 (1992): 19–52.

41. Rollin McCraty et al., "The Electricity of Touch: Detection and Measurement of Cardiac Energy Exchange between People," in *Brain and Values: Is a Biological Science of Values Possible?,* ed. Karl H. Pribram (Lawrence Erlbaum Associates, 1998), 359–79.

42. M. Yamamoto et al., "An Experiment on Remote Action against Man in Sense Shielding Condition," *Journal of the International Society of Life Information Sciences* 14, no. 1 (1996): 97–99.

43. S. Cohen and F.-A. Popp, "Biophoton Emission of the Human Body," *Journal of Photochemistry and Photobiology* 40 (1997): 187–89.

44. F.-A. Popp et al., "Mechanism of Interaction between Electromagnetic Fields and Living Organisms," *Science in China* (Series C) 43, no. 5 (2000): 507–18.

45. Interview with Popp, Neuss, Germany, March 1, 2006.

46. K. Nakamura and M. Hiramatsu, "Ultra-weak Photon Emission from Human Hand: Influence of Temperature and Oxygen Concentration on Emission," *Journal of Photochemistry and Photobiolology B: Biology* 80, no. 2 (2005): 156–60.

47. Hutchison et al., "Pain-related Neurons."

48. "Bees and Ants 'Operate in Teams,'" BBC News web page, March 23, 2009 http://news.bbc.co.uk/1/hi/scotland/7957834.stm

CHAPTER 5

1. Ayn Rand, *The Fountainhead* (Signet, 1996).

2. F. Nietzsche, *Human, All Too Human: Beyond Good and Evil* (Wordsworth Edi-

tions, 2008); F. Nietzsche, *The Gay Science: With a Prelude in Rhymes and an Appendix of Songs* (Vintage, 1974).

3. Robert L. Bear, *Delivered unto Satan* (Bear, 1974).

4. E. Durkheim, *Suicide* (1897; reprinted, Free Press, 1997).

5. M. Daley et al, "Relative Status and Well-being: Evidence from U.S. Suicide Deaths," Federal Reserve Bank of San Francisco Working Paper Series, May 2008, No. 2007–12.

6. A. Clark and C. Senik, "Who Compares to Whom? The Anatomy of Income Comparisons in Europe," IZA Discussion Papers 4414, Institute for the Study of Labor (IZA) (2009).

7. S. L. Syme et al., "Some Social and Cultural Factors Associated with the Occurrence of Coronary Heart Disease," *Journal of Chronic Diseases* 17 (1964): 277–89; S. L. Syme et al., "Cultural Mobility and the Occurrence of Coronary Heart Disease," *Journal of Health and Human Behavior* 6 (1965): 178–89.

8. M. G. Marmot and S. L. Syme, "Acculturation and Coronary Heart Disease in Japanese-Americans," *American Journal of Epidemiology* 104, no. 3 (1976): 225–47; S. L. Syme et al., "Epidemiologic Studies of Coronary Heart Disease and Stroke in Japanese Men Living in Japan, Hawaii and California: Introduction," *American Journal of Epidemiology* 102, no. 6 (1975): 477–80.

9. D. Reed et al., "Social Networks And Coronary Heart Disease Among Japanese Men In Hawaii," *American Journal of Epidemiology* 117 (1983): 384–96.

10. Y. S. Matsumoto, "Social Stress and Coronary Heart Disease in Japan: A Hypothesis," *Milbank Memorial Fund Quarterly* 48 (1970): 9–36, cited in S. Wolf and J. G. Bruhn, *The Power of Clan* (Transaction Publishers, 1993), 92–93.

11. L. Berkman et al., "Social Networks, Host Resistance, and Mortality: A Nine-year Follow-up Study of Alameda County Residents," *American Journal of Epidemiology* 109 (1979): 186–204; *American Journal of Epidemiology* 128 (1988): 370–80.

12. J.K. Vormbrock, and J. M. Grossberg, "Cardiovascular effects of human-pet dog interactions" *Journal of Behavioral Medicine* 11 (1988): 509–17.

13. Wolf and Bruhn, *The Power of Clan.*

14. V. Fuchs, *Who Shall Live?* (Basic Books, 1975), cited in ibid.

15. L. B. Page et al., "Antecedents of Cardiovascular Disease in Six Solomon Island Societies," *Circulation* 49 (1974): 1132–46.

16. M. J. Klag, "The Contribution of Urinary Cations to the Blood Pressure Differences Associated With Migration," *American Journal of Epidemiology* 142, no. 3 (1995): 295–303; J. He, "Effect of Migration and Related Environmental Changes on Serum Lipid Levels in Southwestern Chinese Men," *American Journal of Epidemiology* 144, no. 9 (1996): 839–48; J. He, "Dietary Macronutrients and Blood Pressure in Southwestern China," *Journal of Hypertension* 13, no. 11 (1995): 1267–74; J. He, "Body Mass and Blood Pressure in a Lean Population in Southwestern China," *American Journal of Epidemiology* 139, no. 4 (1994): 380–89.

17. L. F. Berkman and S. L. Syme, "Social Networks, Host Resistance, and Mortality: A Nine-Year Follow-up Study of Alameda County Residents," *American Journal of Epidemiology* 109 no. 2 (1979): 186–204; G. A. Kaplan et al, "Social Connections and Mortality from All Causes and from Cardiovascular Disease: Prospec-

tive Evidence from Eastern Finland," *American Journal of Epidemiology* 128, no. 2 (1988): 370–80.

18. Ibid.

19. B. Boden-Albala et al., "Social Isolation and Outcomes Post Stroke," *Neurology* 64 no. 11 (2005): 1888–92.

20. J. Holt-Lunstad et al., "Social Relationships and Mortality Risk: A Meta-analytic Review," *PLoS Medicine* 7, no. 7 (2010): e1000316.

21. L. Scherwitz et al. "Self-involvement and Coronary Heart Disease Incidence in the Multiple Risk Factor Intervention Trial," *Psychosomatic Medicine* 48 (1986): 187–99.

22. M. P. Aranda, "Relationship Between Religious Involvement and Psychological Well-Being: A Social Justice Perspective," *Health & Social Work* 33, no.1 (2008): 9–21.

23. C. B. Thomas and D. C. Ross, "Precursors of Hypertension and Coronary Disease among Healthy Medical Students: Discriminant Function Analysis V. Family Attitudes," *Johns Hopkins Medical Journal* 123 (1968): 283–96.

24. A. Hawton, "The Impact of Social Isolation on the Health Status and Health-related Quality of Life of Older People," *Quality of Life Research,* July (2010) online: DOI.

25. I. Kawachi et al., "Social Capital, Income Inequality, and Mortality," *American Journal of Public Health* 87, no. 9 (1997): 1491–98.

26. S. Cohen et al., "Social Ties and Susceptibility to the Common Cold," *Journal of the American Medical Association* 277, no. 24 (1997): 1940–44.

27. R. D. Putnam, *Bowling Alone: The Collapse and Revival of American Community* (Simon & Schuster, 2000); R. D. Putnam, "Social Capital: Measurement and Consequences," *ISUMA: Canadian Journal of Policy Research* 2, no. 1 (2001): 41–51.

28. E. Day, "Why Reality TV Works," *The Observer,* Sunday, November 21, 2010.

29. P. G. Zimbardo, *The Lucifer Effect: Understanding How Good People Turn Evil* (Random House, 2007).

CHAPTER 6

1. Stuart Nicolson, "Child Vampire Hunters Sparked Comic Crackdown," BBC News online, March 22, 2010.

2. D. Ellis and D. Anderson, *Conflict Resolution: An Introductory Text* (Emond Montgomery, 2005).

3. "Teenagers Hit by Soap Opera Virus," Reuters, May 19, 2006.

4. B. Appleyard, "Why They Kill Themselves," *Sunday Times,* November 7, 2010.

5. S. Atran, *Talking to the Enemy: Violent Extremism, Sacred Values and What It Means to Be Human* (Allen Lane, 2010).

6. E. Hatfield et al., "Primitive Emotional Contagion: Recent Research," in *The Social Neuroscience of Empathy,* ed. J. Decety and W. Ickes (MIT Press, 2011).

7. www.edu-cyberpg.com/Literacy/whatresearchCondon.asp.

8. William S. Condon, "Cultural Microrhythms," in *Interaction Rhythms: Periodicity in Communicative Behavior,* ed. M. Davis (Human Science Press, 1982), 53–76; William S. Condon, "Method of Micro-analysis of Sound Films of Behavior," *Behavior Research Methods, Instruments and Computers* 2, no. 2 (1970): 51–54.

9. W. S. Condon and L. W. Sander, "Synchrony Demonstrated between Movements of the Neonate and Adult Speech," *Child Development* 45 (1974): 456–62; W. S. Condon and L. W. Sander, "Neonate Movement Is Synchronised with Adult Speech: Interactional Participation and Language Acquisition," *Science* 183 (1974): 99–101.

10. W. S. Condon, "Method of micro-analysis of sound films of behavior." *Behavior Research Methods, Instruments & Computers* 2, no. 2 (1970): 51–54.

11. W. S. Condon and L. W. Sander, "Neonate Movement Is Synchronised with Adult Speech."

12. Cited in E. T. Hall, *Beyond Culture* (Anchor Books, 1989), 74.

13. Ibid., 71.

14. Ibid.

15. Hall, *Beyond Culture,* 76–77.

16. F. de Waal, *The Age of Empathy* (Random House, 2009), 48–49.

17. J. E. Warren et al., "Positive Emotions Preferentially Engage an Auditory-motor Mirror System," *Journal of Neuroscience* 26, no. 50 (2006): 13067–75.

18. De Waal, *The Age of Empathy,* 50–51.

19. Hatfield, et. al, "Primitive Emotional Contagion."

20. M. Karkovsky, "When Feelings Go to Work," *Wharton Alumni Magazine,* spring 2006.

21. S. G. Barsade, "The Ripple Effect: Emotional Contagion and Its Influence on Group Behavior," *Administrative Science Quarterly* 47 (2002): 644–75.

22. Karkovsky, "When Feelings Go to Work."

23. R. Neumann and F. Strack, "'Mood Contagion': The Automatic Transfer of Mood between Persons," *Journal of Personality and Social Psychology* 79 (2000): 211–23.

24. M. Schedlowski et al., "Psychophysiological, Neuroendocrine and Cellular Immune Reactions Under Psychological Stress," *Neuropsychobiology* 28 (1993): 87–90; J. M. Scanlan et al., "CD4 and CD8 Counts Are Associated with Interactions of Gender and Psychosocial Stress," *Psychosomatic Medicine* 60, no. 5 (1998): 644–53.

25. J. K. Kiecolt-Glaser et al., "Negative Behavior During Marital Conflict Is Associated with Immunological Down-regulation," *Psychosomatic Medicine* 55 (1993): 395–409.

26. S. E. Sephton et al., "Diurnal Cortisol Rhythm as a Predictor of Breast Cancer Survival," *Journal of the National Cancer Institute* 92, no. 2 (2000): 994–1000.

27. M. E. J. Newman, "The Structure and Function of Complex Networks," The Society for Industrial and Applied Mathematics *SIAM Review* 45 (2003): 167–256.

28. M. McPherson et al., "Birds of a Feather: Homophily in Social Networks," *Annual Review of Sociology* 27 (2001): 415–44.

29. N. A. Christakis et al., "The Spread of Obesity in a Large Social Network over 32 Years," *New England Journal of Medicine* 357 (2007): 370–79.

30. J. H. Fowler and N. A. Christakis, "Dynamic Spread of Happiness in a Large Social Network: Longitudinal Analysis over 20 Years in the Framingham Heart Study," *British Medical Journal* 337 (2008): a2338.

31. N. Christakis and J. Fowler, *Connected: The Amazing Power of Social Networks and How They Shape Our Lives* (HarperPress, 2010).

NOTES

CHAPTER 7

1. S. Oliner, *Do unto Others: Extraordinary Acts of Ordinary People* (Westview Press, 2003), 13.
2. R. Dawkins, *The Selfish Gene* (Oxford University Press, 1989).
3. Robert Trivers, *Natural Selection and Social Theory: Selected Papers of Robert Trivers,* Evolution and Cognition Series (Oxford University Press, 2002).
4. Oren Solomon Harman, *The Price of Altruism: George Price and the Search for the Origins of Kindness* (W. W. Norton & Company, 2010).
5. Ibid.
6. Dawkins, *The Selfish Gene.*
7. P. A. Kropotkin, *Mutual Aid: A Factor of Evolution* (General Books, 2010).
8. V. H. Cahalane, "Badger-Coyote 'Partnerships,'" *Journal of Mammalogy* 31 (1950): 354–55.
9. G. S. Wilkerson, "Reciprocity and Sharing in the Vampire Bat," *Nature* 308 (1984): 181–84.
10. D. L. Cheney and R. M. Seyfarth, *How Monkeys See the World: Inside the Mind of Another Species* (University of Chicago Press, 1990).
11. H. Kummer, *Social Organization of Hamadryas Baboons: A Field Study* (University of Chicago Press, 1968).
12. B. P. Wiesner and N. M. Sheard, *Maternal Behaviour in the Rat* (Oliver & Boyd, 1933).
13. M. Bekoff and J. Pierce, *Wild Justice: The Moral Lives of Animals* (University of Chicago Press, 2009), 56, 59–60, 125.
14. K. E. Langergraber et al., "The Limited Impact of Kinship on Cooperation in Wild Chimpanzees," *Proceedings of the National Academy of Sciences, USA* 104 (2007): 7786–90.
15. G. E. Rice and P. Gainer, "'Altruism' in the Albino Rat," *Journal of Comparative and Physiological Psychology* 55 (1962): 123–25.
16. J. H. Wechkin et al., "Altruistic Behavior in Rhesus Monkeys," *American Journal of Psychiatry* 121 (1964): 584–85.
17. R. M. Church, "Emotional Reactions of Rats to the Pain of Others," *Journal of Comparative Physiological Psychology* 52 (1959): 132–34.
18. F. Warneken and M. Tomasello, "Altruistic Helping in Human Infants and Young Chimpanzees," *Science* 311 (2006): 1301–303; F. Warneken et al., "Spontaneous Altruism by Chimpanzees and Young Children," *PLoS Biology* 5, no. 7 (2007): 1414–1420.
19. Ibid.
20. J. Greene et al, "An fMRI Investigation of Emotional Engagement in Moral Judgment," *Science* 293, no. 5537 (2001): 2105–8. I am indebted to Dacher Keltner, "The Compassionate Instinct," *Greater Good,* spring 2004, 6–9, for his research into the studies of the biological basis of caring.
21. J. Moll et al. "Human Fronto-mesolimbic Networks Guide Decisions about Charitable Donation," *Proceedings of the National Academy of Sciences of the United States of America* 103 (2006): 15623–28.
22. J. Rilling. et al., "A Neural Basis for Social Cooperation," *Neuron* 35(2) (2002): 395–405.

23. "Emory Brain Imaging Studies Reveal Biological Basis For Human Cooperation," *ScienceDaily* (July 18, 2002) online.

24. Ibid.

25. Keltner, "The Compassionate Instinct."

26. S. G. Post, "Altruism, Happiness and Health; It's Good to Be Good," *International Journal of Behavioral Medicine* 12, no. 2 (2005): 66–77.

27. All of the details of this study have been taken from Joshua Wolf Shenk, "What Makes Us Happy?," *The Atlantic,* June 2009.

28. A good deal of subsequent research has challenged the facts of the Genovese case, according to Jim Rasenberger, "Nightmare on Austin Street, " *American Heritage* October 5, 2006 online; R. Manning et al. "The Kitty Genovese Murder and the Social Psychology of Helping: the Parable of the 38 Witnesses," *American Psychologist* 62 (2007): 555–62.

29. Studies listed in I. M. Piliavin, "Good Samaritanism: An Underground Phenomenon?," *Journal of Personality and Social Psychology* 13, no. 4 (1969): 289–99.

30. Mario Beauregard and Denyse O'Leary, *The Spiritual Brain: A Neuroscientist's Case for the Existence of the Soul* (HarperOne, 2007), 10.

31. I. M. Piliavin, "Good Samaritanism: An Underground Phenomenon?"

32. Oliner, *Do unto Others.*

33. Ibid., 45.

34. Svetlana Broz, "Courage under Fire," *Greater Good* 3 (fall/winter 2006–7): 10–13.

35. Ibid.; Jason Marsh, "The Making of an Upstander," *Greater Good* 3 (fall/winter 2006–7): 10–13.

36. M. T. Ghiselin, *The Economy of Nature and the Evolution of Sex* (University of California Press, 1974), 247, cited in H. Gintis et al., "Explaining Altruistic Behavior in Humans," *Evolution and Human Behavior* 24 (2003): 153–72.

37. C. D. Batson et al., "Five Studies Testing Two New Egoistic Alternatives to the Empathy-Altruism Hypothesis," *Journal of Personality and Social Psychology* 55 (1988): 52–77.

38. C. Lamm et al., "The Neural Substrate of Human Empathy: Effects of Perspective-taking and Cognitive Appraisal," *Journal of Cognitive Neuroscience* 19 (2007): 42–58.

39. C. D. Batson et al., "Perspective Taking: Imagining How Another Feels versus Imagining How You Would Feel," *Personality and Social Personality Bulletin* 23 (1997): 751–58.

40. R. B. Cialdini et al., "Reinterpreting the Empathy-Altruism Relationship: When One into One Equals Oneness," *Journal of Personality and Social Psychology* 73, no. 93 (1997): 481–94.

CHAPTER 8

1. M. Spencer, "Rapoport at Ninety," *Connections* 24, no. 3 (2002): 104–107.

2. The best known is R. Axelrod, *The Evolution of Cooperation* (Basic Books, 1984).

3. A. M. Colman and L. Browning, "Evolution of Cooperative Turn-taking," *Evolutionary Ecology Research* 11 (2009): 949–63.

4. H. Gintis et al., "Explaining Altruistic Behavior in Humans," *Evolution and Human Behavior* 24 (2003): 153–72.

5. J. Henrich et al., "Overview and Synthesis," in *Foundations of Human Sociality*, ed. J. Henrich et al. (Oxford University Press, 2004), cited in R. Wilkinson and Kate Pickett, *The Spirit Level: Why More Equal Societies Almost Always Do Better* (Allen Lane/Penguin, 2009), 199.

6. Ibid.

7. J. Henrich et al., "Cooperation, Reciprocity and Punishment in Fifteen Small-scale Societies," *American Economic Review* 91 (2001): 73–78.

8. S. Brosnan and B. M. De Waal, "Animal Behaviour: Fair Refusal by Capuchin Monkeys," *Nature* 428 (2004): 140.

9. S. Markey, "Monkeys Show Sense of Fairness, Study Says," *National Geographic News*, September 17, 2003, http://news.nationalgeographic.com/news/2003/09/0917_030917_monkeyfairness.html.

10. See also K. Jensen, "Chimpanzees Are Rational Maximizers in an Ultimatum Game," *Science* 318, no. 5847 (2007): 107–9.

11. E. Tricomi et al, "Neural Evidence for Inequality-averse Social Preferences," *Nature* 463 (2010): 1089–91.

12. E. Fehr et al., "Reciprocity as a Contract Enforcement Device: Experimental Evidence," *Econometrica* 65 (1997): 437–59.

13. T. Baumgartner, "Oxytocin Shapes the Neural Circuitry of Trust and Trust Adaptation in Humans," *Neuron* 58 (2008): 639–50.

14. C. Eisenegger et al., "Prejudice and Truth about the Effect of Testosterone on Human Bargaining Behaviour," *Nature* 463 (2010): 356–59.

15. Interview with E. Fehr, February 15, 2010.

16. E. Fehr and K. M. Schmidt, "A Theory of Fairness, Competition and Cooperation," *Quarterly Journal of Economics*, 114 (1999): 817–68.

17. T. Singer et al., "Empathic Neural Responses Are Modulated by the Perceived Fairness of Others," *Nature* 439 (2006): 466–69.

18. R. Wilkinson and Kate Pickett, *The Spirit Level*.

19. Ibid.

20. R. McCarthy and G. Kiser, "Number of U. S. Millionaires Soared in 2009: Spectrem Group," Huffington Post, May 9, 2010.

21. Stephen Wright, "Fat Cats in Terror After Anti-capitalists Attack Fred the Shred's Home," Daily Mail online, March 26, 2009. http://www.dailymail.co.uk/news/article-1164691/Fat-cats-terror-anti-capitalists-attack-Fred-Shreds-home.html#ixzz18xks0rmr.

22. R. Putnam et al., "The Social Capital Community Benchmark Survey," http://www.hks.harvard.edu/saguaro/communitysurvey/results_pr.html.

23. Eric Uslaner, *The Moral Foundations of Trust* (Cambridge University Press, 2002).

24. J. Lanzetta and Basil Englis, "Expectations of Cooperation and Competition and Their Effects on Observers' Vicarious Emotional Responses," *Journal of Personality and Social Psychology* 56, no. 4 (1989): 543–54.

25. I am indebted to E. L. Doctorow, who used this phrase in a review of J. Johnson, *Bad Connections* (Pocket, 1979).

NOTES

CHAPTER 9

1. Neelesh Misra, "Stone Age Cultures Survive Tsunami Waves," Associated Press, January 4, 2005.
2. A. Gislén, "Superior Underwater Vision in a Human Population of Sea Gypsies," *Current Biology* 13, no. 10 (2003): 833–36.
3. "The Sea Gypsies," *60 Minutes,* March 20, 2005.
4. Amelia Gentleman, "Smile Please," *The Guardian,* October 19, 2004.
5. Richard Nisbett, *The Geography of Thought: How Asians and Westerners See Things Differently and Why* (Free Press, 2003).
6. Ibid., 50.
7. D. Monro, ed., *Individualism and Holism: Studies in Confucian and Taoist Values* (Center for Chinese Studies, University of Michigan, Ann Arbor, 1985), 1–34, cited in R. Nisbett and T. Masuda, "Culture and Point of View," *Proceedings of the National Academy of Sciences of the United States of America* 100, no. 19 (2003): 11163–70.
8. D. L. Fixico, *The American Indian Mind in a Linear World* (Routledge, 2003), 3.
9. G. V. Goddard, "Development of Epileptic Seizures through Brain Stimulation at Low Intensity," *Nature* 214 (1967): 1020.
10. R. M. Post, "Transduction of Psychosocial Stress into the Neurobiology of Recurrent Affective Disorder," *American Journal of Psychiatry* 149 (1992): 999–1010; J. Ormel et al., "Vulnerability before, during, and after a Major Depressive Episode: A 3-Wave Population-Based Study," *Archives of General Psychiatry* 61 (2004): 990–96; K. P. Bailey, "Electrophysiological Kindling and Behavioral Sensitization as Models for Bipolar Illness: Implications for Nursing Practice," *Journal of the American Psychiatric Nurses Association* 5 (1999): 62–66.
11. Nisbett and Masuda, "Culture and Point of View."
12. T. Masuda and R. E. Nisbett, "Attending Holistically vs. Analytically: Comparing the Context Sensitivity of Japanese and Americans," *Journal of Personality and Social Psychology* 81 (2001): 922–34.
13. Nisbett and Masuda, "Culture and Point of View."
14. H. F. Chua et al., "Cultural Variation in Eye Movements during Scene Perception," *Proceedings of the National Academy of Science of the United States of America* 102, no. 35 (2005): 12629–33.
15. Daniel J. Simons, a psychologist at the University of Illinois at Urbana-Champaign, first created this study, which Brown amplified. See J. Simons and D. T. Levin, "Failure to Detect Changes to People During a Real-world Interaction," *Psychonomic Bulletin and Review* 5 (1998): 644–49. Derren Brown Person Swap, available online as "Mind Control," S01E02, "Trick of the Mind," S03E04.
16. T. Grandin, *Animals in Translation* (Scribner, 2005).
17. www.aquaticsintl.com/2004/nov/0411_rm.html.
18. A. Mack and I. Rock, "Inattentional Blindness: An Overview," *Current Directions in Psychological Science* 12, no. 5 (2003): 180–84.
19. I. Baruch et al., "Latent Inhibition and 'Psychotic Proneness' in Normal Subjects," *Personality and Individual Differences* 9 (1988): 777–83.
20. S. Carson, "Decreased Latent Inhibition Is Associated with Increased Creative

Achievement in High-Functioning Individuals," *Journal of Personality and Social Psychology* 85, No. 3 (2003): 499–506.

21. A. Gislén et al., "Visual Training Improves Underwater Vision in Children," *Vision Research* 46, no. 20 (2006): 3443–50.

22. W. H. Goodenough and S. D. Thomas, "Traditional Navigation in the Western Pacific: A Search for Pattern," *Expedition* 29, no. 3 (1987): 3–14.

23. Grandin, *Animals in Translation.*

24. Ibid.

25. A. W. Snyder et al., "Savant-like Skills Exposed in Normal People by Suppressing the Left Fronto-temporal Lobe," *Journal of Integrative Neuroscience* 2, no. 2 (2003): 149–58.

26. J. Panksepp, "The Anatomy of Emotions," in *Emotion: Theory, Research and Experience, Vol. 3: Biological Foundations of Emotions,* ed. R. Plutchik (Academic Press, 1986), 91–124.

27. Ibid.

28. G. E. Swan and D. Carmelli, "Curiosity and Mortality In Aging Adults: A 5-Year Follow-Up of the Western Collaborative Group Study," *Psychology and Aging* 11 (1996): 449–53.

29. W. Woodward, "New Surprises in Very Old Places: Civil War Nurse Leaders and Longevity," *Nursing Forum* 26, no. 1 (1991): 9–16.

30. J. Panksepp, "The Neurobiology of Emotions: Of Animal Brains and Human Feelings," in *Handbook of Psychophysiology,* ed. T. Manstead and H. Wagner (John Wiley & Sons, 1989), 5–26.

31. B.R. Dunn et al., "Concentration and Mindfulness Meditations: Unique Forms of Consciousness?" *Applied Psychophysiology and Biofeedback* 24, no. 3 (1999): 147–65.

32. D. Brown et al., "Visual Sensitivity and Mindfulness Meditation," *Perceptual and Motor Skills* 58, no. 3 (1984): 727–33, 775–84.

33. A. Lutz et al., "Attention Regulation and Monitoring in Meditation," *Trends in Cognitive Science* 12, no. 4 (2008): 163–69.

34. H. A. Slagter et al., "Mental Training Affects Distribution of Limited Brain Resources," *PLoS Biology* 5, no. 6 (2007): e138.

35. Lutz et al., "Attention Regulation."

36. J. Reiman, "The Impact of Meditative Attentional Training on Measures of Select Attentional Parameters and on Measures of Client-Perceived Counsellor Empathy," *Dissertation Abstracts International* 46, no. 6A (1985): 1569. See also S. Donovan et al., *The Physical and Psychological Effects of Meditation,* chapter 3 (Institute of Noetic Sciences, 1997).

37. For this and other quotes of Don Beck, Interview with Beck, October 19, 2010.

38. M. Gerzon, *Global Citizens* (Harvard Business School Press, 2006), 180.

39. Ibid.

40. Ibid., 33.

CHAPTER 10

1. For this and all quotations and information about Orland Bishop, interviews, April 16 and October 16, 2010.

NOTES

2. M. Buber, *I and Thou* (Continuum, 2004).

3. S. D. Cousins, "Culture and self-perception in Japan and the United States," *Journal of Personality and Social Psychology* 56 (1989): 124–31; C. Kanagawa et al., "'Who Am I?' The Cultural Psychology of the Conceptual 'Self,'" *Personality and Social Psychology Bulletin* 27 (2001): 90–103, cited in R.E. Nisbett and T. Masuda, "Culture and Point of View," *Proceedings of the National Academy of Sciences of the United States of America* 100, no. 19 (2003): 11163–70.

4. H. Kim and H. R. Markus, "Deviance or uniqueness, harmony or conformity?: A cultural analysis," *Journal of Personality and Social Psychology* 77 (1999): 785–800, cited in R. E. Nisbett and T. Masuda, "Culture and Point of View."

5. M. McPherson et al., "Birds of a Feather: Homophily in Social Networks," *Annual Review of Sociology* 27 (2001): 415–44.

6. S. G. Barsade and D. E. Gibson, "Group Emotion: A View from Top and Bottom," *Research on Managing Groups and Teams* 1 (1998): 81–102.

7. D. Bohm et al., "Dialogue: A Proposal." Available online at www.muc.de/~heuvel/dialogue/dialogue_proposal.html.

8. Ibid.

9. Ibid.

10. "Public Conversations Project, 'Talking with the Enemy,'" *Boston Globe,* January 28, 2001.

11. Interview with James O'Dea, October 9, 2010.

12. A. Dillard, "Living Like Weasels," in *Teaching a Stone to Talk* (Harper Colophon, 1982), 13–14.

13. T. Lewis et al., *A General Theory of Love* (First Vintage, 2000), 63–64.

14. J. Grinberg-Zylberbaum et al., "The Einstein-Podolsky-Rosen Paradox in the Brain: The Transferred Potential," *Physics Essays* 7, no. 4 (1994): 422–28.

15. Rollin McCraty et al., "The Electricity of Touch: Detection and Measurement of Cardiac Energy Exchange between People," in *Brain and Values: Is a Biological Science of Values Possible?*, ed. Karl H. Pribram (Lawrence Erlbaum Associates, 1998), 359–79.

16. R. J. Davidson and W. Irwin, "The Functional Neuroanatomy of Emotion and Affective Style," *Trends in Cognitive Sciences* 3 (1999): 11–21.

17. A. Lutz et al., "Regulation of the Neural Circuitry of Emotion by Compassion Meditation: Effects of Meditative Expertise," *PLoS ONE* 3, no. 3 (2008): e1897, online.

18. A. B. Newberg and J. Iversen, "The Neural Basis of the Complex Mental Task of Meditation: Neurotransmitter and Neurochemical Considerations," *Medical Hypotheses* 61, no. 2 (2003): 282–91.

19. G. Müller-Fahrenholz, *The Art of Forgiveness: Theological Reflections on Healing and Reconciliation* (WCC Publication, 1997).

20. Ibid.

21. G. Müller-Fahrenholz, "On Shame and Hurt in the Life of Nations: A German Perspective," *Irish Quarterly Review* 78 (1989): 127–35, cited in J. Thompson, "On Forgiveness and Social Healing," Panel Discussion on the Role of Forgiveness in *Social Healing, Harvard Divinity School, October 31, 2005.* Available online at www.humiliationstudies.org/documents/ThompsonForgiveness.pdf.

22. J. Thompson, "On Forgiveness and Social Healing."

23. Ibid.
24. Ibid.; and interview with O'Dea, op cit.
25. Ibid.
26. G. Müller-Fahrenholz, *The Art of Forgiveness.*
27. For all evidence to date about The Intention Experiments, please see www.theintentionexperiment.com.
28. A. Seeger, *Why Suyá Sing* (University of Illinois Press, 2004).
29. A. Bellos, *Here's Looking at Euclid: A Surprising Excursion Through the Astonishing World of Math* (Free Press, 2010).

CHAPTER 11

1. Richard Szymanski, "Can Changing Neighborhood Traffic Circulation Patterns Reduce Crime and Improve Personal Safety? A Quantitative Analysis of One Neighborhood's Efforts," unpublished master's thesis, Florida Atlantic University, Department of City and Regional Planning, 1994. The author was former captain of the Fort Lauderdale Police Department.
2. D. Diamond, "Behind Closed Gates," *USA Today,* 1997, 1: 1–3, as quoted in E. J. Drew and J. M. McGuigan, "Prevention of Crime: An Overview of Gated Communities and Neighborhood Watch," International Foundation for Protection Officers, online: ttp://www.ifpo.org/articlebank/gatedcommunity.html.
3. U.S. Department of Justice, Bureau of Justice Statistics, *National Crime Victimization Survey,* 1993, cited in E. J. Blakely and M. G. Snyder, "Separate Places: Crime and Security in Gated Communities," in *Reducing Crime through Real Estate Development and Management,* ed. M. Fleson and R. B. Peiser (Urban Land Institute, 1998), 53–70.
4. G. S. Rudoff, "Cell Phones Reduce Major Urban Crimes," American City and County 111 (1996): 19–20, as reported in Drew and McGuigan, "Prevention of Crime."
5. Edward J. Blakely and Mary Gail Snyder, *Fortress America: Gated Communities in the United States* (The Brookings Institution, 1999).
6. R. D. Putnam, "E Pluribus Unum: Diversity and Community in the Twenty-first Century: The 2006 Johan Skytte Prize Lecture," *Scandinavian Political Studies* 30, no. 2 (2007): 137–74.
7. M. Sherif et al., *The Robbers Cave Experiment: Intergroup Conflict and Cooperation, 1954* (University of Oklahoma Institute of Group Relations, 1961).
8. G. W. Allport, *The Nature of Prejudice* (Perseus Books, 1954); T. F. Pettigrew, "Intergroup Contact Theory," *Annual Review of Psychology* 49 (1998): 65–85.
9. D. Bramel, "The Strange Career of the Contact Hypothesis," in *The Psychology of Ethnic and Cultural Conflict,* ed. Y. T. Lee, C. McCauley, F. Moghaddam, and S. Worchel (Praeger, 2004), 49–69.
10. T. F. Pettigrew and L. R. Tropp, "A Meta-analytic Test of Intergroup Contact Theory," *Journal of Personality and Social Psychology* 90, no. 5 (2006): 751–83.
11. Interview with Don Beck, October 19, 2010.
12. R. I. Sutton, "Boss Luis Urzúa and the Trapped Miners in Chile: A Classic Case of Leadership, Performance, and Humanity," *Psychology Today,* September 6, 2010; J. Webber and J. P. Rathbone, "Man in the News: Luis Urzúa," *Financial*

NOTES

Times, London, October 15, 2010; Nick Kanas, "Notes for the Underground," *New York Times* OpEd, August 29, 2010, online.

13. U. Lindenberger et al., "Brains Swinging in Concert: Cortical Phase Synchronization while Playing Guitar," *BMC Neuroscience* 10 (2009): 22.

14. E. A. Cohen et al., "Rowers' High: Behavioural Synchrony is Correlated with Elevated Thresholds," *Biology Letters* 6 (2010): 106–8.

15. "Tailholt Residents Build Community Center," *Cherokee News Path,* March 10, 2005, online.

16. D. Agent, "Tailholt Community Reaps Rewards of Gadugi," *Cherokee Phoenix,* Tahlequah, Oklahoma, March 2006.

17. Interview with Yoko Masashiro via Fukiko Kai, translator, May 26, 2010.

18. S. Wolf and J. G. Bruhn, *The Power of Clan.*

19. Interview with Nour Hakki, April 20, 2010.

20. J. C. Deschamps and W. Doise, "Crossed Category Memberships in Intergroup Relations," In H. Tajfel (ed.), *Differentiation Between Social Groups* (Cambridge University Press, 1978), 141–58.

21. J. Jetten et al, " The Social Cure?" *Scientific American Mind,* September/October 2009.

22. H. Tajfel, "Experiments in Intergroup Discrimination," *Scientific American* 223 (1970): 96–102.

23. R. D. Putnam and David E. Campbell, *American Grace: How Religion Divides and Unites Us* (Simon & Schuster, 2010).

CHAPTER 12

1. "Marie's" identity has been changed.

2. H. Gintis et al., "Explaining Altruistic Behavior in Humans," *Evolution and Human Behavior* 24 (2003), 153–72.

3. N. Christakis, email correspondence with author, February 19, 2010.

4. J. Fowler and N. Christakis, "Cooperative Behavior Cascades in Human Social Networks," *Proceedings of the National Academy of Sciences of the United States of America* 107, no. 12 (2010): 5334–38.

5. J. S. Lewis, "Dear to My Heart," speech recorded at the BBC, April 15, 1957.

6. www.hotdogonastick.com.

7. Robert D. Putnam and Lewis M. Feldstein, *Better Together: Restoring the American Community* (Simon & Schuster, 2003).

CHAPTER 13

1. David W. Johnson and Roger T. Johnson, "An Educational Psychology Success Story: Social Interdependence Theory and Cooperative Learning," *Educational Researcher* 38 (2009): 365–79.

2. Interview with Don Beck, October 19, 2010.

3. Interview with Jack Canfield, July 12, 2011.

4. E. E. S. Bjornstrom, "The Neighborhood Context of Relative Position, Trust, and Self-Rated Health," *Social Science & Medicine* 73 (2011): 42–49.

BIBLIOGRAPHY

Abbott, A. "Testosterone Link to Aggression May Be All in the Mind." *Nature News,* December 8, 2009. Online.

"After You: How Taking Turns Is Hardwired into Our Evolution." *Mail Online,* July 9, 2009.

Agent, D. "Tailholt Community Reaps Rewards of Gadugi." *Cherokee Phoenix* (Tahlequah, Oklahoma), March 2006.

Allahverdiyev, A. R., et al. "Possible Space Weather Influence on Functional Activity of the Human Brain." Presentation at the Space Weather Workshop: Looking towards a European Space Weather Programme. Noordwijk, The Netherlands, December 17–19, 2001.

Alleyne, R. "Britain's 'Me Culture' Making Us Depressed." *Telegraph,* November 6, 2009.

Allport, G. W. *The Nature of Prejudice.* Perseus Books, 1954.

Alonso, Y. "Geophysical Variables and Behavior: LXXII. Barometric Pressure, Lunar Cycle, and Traffic Accidents." *Perceptual and Motor Skills* 77, no. 2 (1993): 371–76.

Alves, D. W., et al. "Effect of Lunar Cycle on Temporal Variation in Cardiopulmonary Arrest in Seven Emergency Departments During 11 Years." *European Journal of Emergency Medicine* 10, no. 3 (2003): 225–28.

Arai, J., et al. "Transgenerational Rescue of a Genetic Defect in Long-Term Potentiation and Memory Formation by Juvenile Enrichment." *Journal of Neuroscience* 29, no. 5 (2009): 1496–502.

Aranda, M. P. "Relationship Between Religious Involvement and Psychological Well-Being: A Social Justice Perspective." *Health and Social Work* 33, no. 1 (2008): 9–21.

Aspelmeyer, M., and A. Zeilinger. "A Quantum Renaissance." physicsworld.com, July 1, 2008.

Associated Press. "Ancient Tribe Survives Tsunami." January 6, 2005.

———. "Ancient Tribes Touched by Tsunami." January 14, 2005.

———. "Fate of Indigenous Tribes Unknown." January 4, 2005.

———. "'Sea Gypsies' Tsunami Rebuild Curse." January 14, 2005.

"The Authentic Expression of My Voice." Video. Global Oneness Project. September 27, 2006.

Babayev, E. "Some Results of Investigations on the Space Weather Influence on Functioning of Several Engineering-Technical and Communication Systems and

Human Health." *Astronomical and Astrophysical Transactions* 22, no. 6 (2003): 861–67.

Baevsky, R. M., et al. "Meta-analyzed Heart Rate Variability, Exposure to Geomagnetic Storms, and the Risk of Ischemic Heart Disease." *Scripta Medica* (Brno) 70, nos. 4–5 (1997): 201–6.

Bailey, K. P. "Electrophysiological Kindling and Behavioral Sensitization as Models for Bipolar Illness: Implications for Nursing Practice." *Journal of the American Psychiatric Nurses Association* 5 (1999): 62–66.

Bair, C. C. "The Heart Field Effect: Synchronization of Healer-Subject Heart Rates in Energy Therapy." *Advances in Body-Mind Medicine* 23, no. 4 (2008–9): 10–19.

Barber, N. *Kindness in a Cruel World: The Evolution of Altruism.* Prometheus Books, 2004.

"Barrio De Paz." Video. Global Oneness Project. February 12, 2007.

Barsade, S. G. "The Ripple Effect: Emotional Contagion and Its Influence on Group Behavior." *Administrative Science Quarterly* 47 (2002): 644–75.

Barsade, S. G., and D. E. Gibson. "Group Emotion: A View from Top and Bottom." *Research on Managing Groups and Teams* 1 (1998): 81–102.

Barsade, Sigal G., and Gibson, Donald E. "Why Does Affect Matter in Organizations?" *Academy of Management Perspectives,* February 2007, 36–59.

Barsade, Sigal G., et al. "The Affective Revolution in Organizational Behavior: The Emergence of a Paradigm." in J. Greenberg, ed., *Organizational Behavior: The State of the Science,* 3–52. Erlbaum Associates, 2003.

Baruch, I., et al. "Latent Inhibition and 'Psychotic Proneness' in Normal Subjects." *Personality and Individual Differences* 9 (1988): 777–83.

Batson, C. D. "How Social an Animal? The Human Capacity for Caring." *American Psychologist* 45, no. 3 (1990): 336–46.

Batson, C. Daniel, and Nadia Ahmad. "Four Motives for Community Involvement." *Journal of Social Issues* 58, no. 3 (2002): 429–45.

Batson, C. D., et al. "Empathic Joy and the Empathy-Altruism Hypothesis." *Journal of Personality and Social Psychology* 61, no. 3 (1991): 413–26.

Batson, C. D., et al. "Five Studies Testing Two New Egoistic Alternatives to the Empathy-Altruism Hypothesis." *Journal of Personality and Social Psychology* 55 (1988): 52–77.

Batson, C. D., et al. "Is Empathic Emotion a Source of Altruistic Motivation?" *Journal of Personality and Social Psychology* 40, no. 2 (1981): 290–302.

Batson, C. D., et al. "Negative-State Relief and the Empathy-Altruism Hypothesis." *Journal of Personality and Social Psychology* 56 (1989): 922–33.

Batson, C. D., et al. "Perspective Taking: Imagining How Another Feels versus Imagining How You Would Feel." *Personality and Social Personality Bulletin* 23 (1997): 751–58.

Batty, David. "Dog Adopts Tiger Triplets." *Guardian,* May 18, 2007. Online.

Baumgartner, T. "Oxytocin Shapes the Neural Circuitry of Trust and Trust Adaptation in Humans." *Neuron* 58 (2008): 639–50.

Beardsley. "In Focus: Evolution Evolving." *Scientific American,* September 1997, 15–16.

Beauregard, Mario, and Denyse O'Leary. *The Spiritual Brain: A Neuroscientist's Case for the Existence of the Soul.* HarperOne, 2007.

BIBLIOGRAPHY

Beck, Don. "Six Games to Glory: A Winning Strategy for the 1995 South African Springbok Team." Unpublished manuscript, 1995.

———. "World Cup Offers Insight into Rugby." *Dallas Morning News*, March 16, 1995.

———. "World Cup Team New Symbol of Unity in South Africa." *Dallas Morning News*, June 8, 1995.

Beck, Don Edward, and Christopher C. Cowan. *Spiral Dynamics: Mastering Values, Leadership, and Change*. Blackwell, 2006.

Becker, Robert O. *The Body Electric: Electromagnetism and the Foundation of Life*. Quill, 1985.

"Bees and Ants 'Operate in Teams.'" *BBC News*. March 23, 2009.

Bellos, A. *Here's Looking at Euclid: A Surprising Excursion Through the Astonishing World of Math*. Free Press, 2010.

Berk, M., et al. "Do Ambient Electromagnetic Fields Affect Behavior? A Demonstration of the Relationship Between Geomagnetic Storm Activity and Suicide." *Bioelectromagnetics* 27 (2006): 151–55.

Berkman, Lisa F., and S. L. Syme. "Social Networks, Host Resistance, and Mortality: A Nine-Year Follow-up Study of Alameda County Residents." *American Journal of Epidemiology* 109, no. 2 (1979): 186–204.

Bernhard, H., et al. "Parochial Altruism in Humans." *Nature* 442 (2006): 912–15.

Bhattacharjee, C., et al. "Do Animals Bite More During a Full Moon? Retrospective Observational Analysis." *British Medical Journal* 321, no. 7376 (2000): 1559–61.

Bishop, Orland. "Keynote Address." Connectivity, São Paulo, Brazil, July 16–22, 2006. Online.

Bjornstrom, E. E. S. "The Neighborhood Context of Relative Position, Trust, and Self-Rated Health." *Social Science & Medicine* 73 (2011): 42–49

Blakely, Edward J., and Mary Gail Snyder. *Fortress America: Gated Communities in the United States*. Brookings Institution, 1999.

Blakely, E. J., and M. G. Snyder. "Separate Places: Crime and Security in Gated Communities." in M. Fleson and R. B., Peiser, eds., *Reducing Crime Through Real Estate Development and Management*, 53–70. Urban Land Institute, 1998.

Blakeslee, S. "A Pregnant Mother's Diet May Turn the Genes Around." *New York Times*, October 7, 2003. Online.

Block-Lerner, J., et al. "The Case of Mindfulness-Based Approaches in the Cultivation of Empathy: Does Nonjudgmental, Present-Moment Awareness Increase Capacity for Perspective-Taking and Empathic Concern?" *Journal of Marital and Family Therapy* 33, no. 4 (2007): 501–16.

Boden-Albala, B., et al. "Social Isolation and Outcomes Post Stroke." *Neurology* 64, no. 11 (2005): 1888–92.

Bohm, D. et al. "Dialogue: A Proposal." www.muc.de/~heuvel/dialogue/dialogue_proposal.html.

Borris, Eileen R. "The Healing Power of Forgiveness." Occasional Paper No 10, Institute for Multi-Track Diplomacy (Washington, D.C.), October 2003.

Bramel, D. "The Strange Career of the Contact Hypothesis." in Y. T. Lee et al., eds., *The Psychology of Ethnic and Cultural Conflict*, 49–69. Praeger, 2004.

Brooks, David. "They Had It Made." Op-ed. *New York Times*, May 12, 2009. Online.

BIBLIOGRAPHY

Brosnan, S., and B. M. De Waal. "Animal Behaviour: Fair Refusal by Capuchin Monkeys." *Nature* 428 (2004): 140.

Brown, D., et al. "Visual Sensitivity and Mindfulness Meditation." *Perceptual and Motor Skills* 58, no. 3 (1984): 727–33, 775–84.

Broz, Svetlana. "Courage under Fire." *Greater Good* 3 (2006–7). Online.

Bruhn, J. G., and S. Wolf. "Studies Reporting 'Low Rates' of Ischemic Heart Disease: A Critical Review." *American Journal of Public Health* 60, no. 8 (1970): 1477–95.

Bruhn, J. G., et al. "Social Aspects of Coronary Heart Disease in Two Adjacent, Ethnically Different Communities." *American Journal of Public Health* 56, no. 9 (1966): 1493–506.

Buber, Martin. *I and Thou*. Continuum, 2004.

Buchenau, J. "Global Darwin: Multicultural Mergers." *Nature* 462, no. 19 (2009): 284–85.

Buettner, Dan. *The Blue Zones: Lessons for Living Longer from the People Who've Lived the Longest*. National Geographic Society, 2008.

Bureau, Y., and M. Persinger. "Decreased Latencies for Limbic Seizures Induced in Rats by Lithium-Pilocarpine Occur when Daily Average Geomagnetic Activity Exceeds 20 Nanotesla." *Neuroscience Letters* 192 (1995): 142–44.

Cairns, J. "The Origin of Mutants." Nature 335 (1988): 142–45.

Campbell, Joseph. *The Hero with a Thousand Faces*. Princeton University Press, 2004.

Carlson, Emily. "How Do Mothers Process Emotion About Their Own Babies?" *Wisconsin Week*, February 27, 2004. Online.

Carson, S. "Decreased Latent Inhibition Is Associated with Increased Creative Achievement in High-Functioning Individuals." *Journal of Personality and Social Psychology* 85, no. 3 (2003): 499–506.

Catmur, C., et al. "Sensorimotor Learning Configures the Human Mirror System." *Current Biology* 17 (2007): 1527–31.

Cavendish, C. "The Big Society Is About Bowling Together." *London Times*, September 23, 2010. Online.

"The Cell: Unraveling the Double Helix and the Secret of Life." *Time*, April 19, 1971. Online.

Chapman, S., and S. Morrell. "Barking Mad? Another Lunatic Hypothesis Bites the Dust." *British Medical Journal* 321, no. 7376 (2000): 1561–63.

Chiao, J. Y., and K. D. Blizinsky. "Culture-Gene Coevolution of Individualism—Collectivism and the Serotonin Transporter Gene." *Proceedings of the Royal Society B: Biological Sciences* 277, no. 1681 (2010): 529–37.

"Child Abuse May 'Mark' Genes in Brains of Suicide Victims." *ScienceDaily*, May 7, 2008.

Chiltern Pierce, J. *Magical Child*. Bantam Books, 1980.

Christakis, N., and J. Fowler. *Connected: The Amazing Power of Social Networks and How They Shape Our Lives*. HarperPress, 2010.

Christakis, N. A., et al. "The Spread of Obesity in a Large Social Network over 32 Years." *New England Journal of Medicine* 357 (2007): 370–79.

Christian, L. M., et al. "Stress and Wound Healing." *Neuroimmunomodulation* 13 (2006): 337–46.

Chua, H. F., et al. "Cultural Variation in Eye Movements During Scene Perception." *Proceedings of the National Academy of Science of the United States of America* 102, no. 35 (2005): 12629–33.

Church, R. M. "Emotional Reactions of Rats to the Pain of Others." *Journal of Comparative Physiological Psychology* 52 (1959): 132–34.

Cialdini, R. B., et al. "Empathy-Based Helping: Is It Selflessly or Selfishly Motivated?" *Journal of Personality and Social Psychology* 52, no. 4 (1987): 749–58.

Cialdini, R. B., et al. "Reinterpreting the Empathy-Altruism Relationship: When One into One Equals Oneness." *Journal of Personality and Social Psychology* 73, no. 93 (1997): 481–94.

Clark, A., and C. Senik. "Who Compares to Whom? The Anatomy of Income Comparisons in Europe." IZA Discussion Papers 4414, Institute for the Study of Labor (Bonn, Germany), 2009.

Climent, C. E., and R. Plutchik. "Lunar Madness: An Empirical Study." *Comprehensive Psychiatry* 18, no. 4 (1977): 369–74.

Cloud, John. "Why Your DNA Isn't Your Destiny." *Time*, January 6, 2010. Online.

Cohen, E. A., et al. "Rowers' High: Behavioural Synchrony Is Correlated with Elevated Thresholds." *Biology Letters* 6 (2010): 106–8.

Cohen, S., and F.-A. Popp. "Biophoton Emission of the Human Body." *Journal of Photochemistry and Photobiology* 40 (1997): 187–89.

Cohen, S., et al. "Social Integration and Health: The Case of the Common Cold." *Journal of Social Structure* 1, no. 3 (2000): 1–7.

Cohen, S., et al. "Social Ties and Susceptibility to the Common Cold." *Journal of the American Medical Association* 277, no. 24 (1997): 1940–44.

Cohen-Mansfield, J., et al. "Full Moon: Does It Influence Agitated Nursing Home Residents?" *Journal of Clinical Psychology* 45, no. 4 (1989): 611–14.

Colman, A. M. "Love Is Not Enough: Other-Regarding Preferences Cannot Explain Payoff Dominance in Game Theory." *Behavioral and Brain Sciences* 30, no. 1 (2007): 22–23.

Colman, A. M., and L. Browning. "Evolution of Cooperative Turn-Taking." *Evolutionary Ecology Research* 11 (2009): 949–63.

Condon, W. S. "Method of Micro-analysis of Sound Films of Behavior." *Behavior Research Methods, Instruments and Computers* 2, no. 2 (1970): 51–54.

Condon, W. S., and L. W. Sander. "Neonate Movement Is Synchronised with Adult Speech: Interactional Participation and Language Acquisition." *Science* 183 (1974): 99–101.

Condon, W. S., and L. W. Sander. "Synchrony Demonstrated Between Movements of the Neonate and Adult Speech." *Child Development* 45 (1974): 456–62.

Cornélissen, Germaine, et al. "Chronomes, Time Structures, for Chronobioengineering for 'a Full Life.'" *Biomedical Instrumentation and Technology* 33, no. 2 (1999): 152–87.

Cornélissen, G., et al. "From Various Kinds of Heart Rate Variability to Chronocardiology." *American Journal of Cardiology* 66 (1990): 863–68.

Cornélissen, G., et al. "Is a Birth-Month-Dependence of Human Longevity Influenced by Half-Yearly Changes in Geomagnetics?" *Proceedings of the XXV Annual Seminar, "Physics of Auroral Phenomena," Apatity* [2657b] (2002): 161–65.

Cornélissen, G., et al. "Non-photic Solar Associations of Heart Rate Variability and Myocardial Infarction." *Journal of Atmospheric and Solar-Terrestrial Physics* 64 (2002): 707–20.

Crisp, R. J., and M. J. Farr. "Moderation of Intergroup Memory Bias via Crossed Categorization." *Current Research in Social Psychology* 9, no. 9 (2004). Online.

BIBLIOGRAPHY

Dalai Lama and Howard C. Cutler. *The Art of Happiness in a Troubled World*. Hodder and Stoughton, 2009.

Daley, M., et al. "Relative Status and Well-being: Evidence from U.S. Suicide Deaths." *Federal Reserve Bank of San Francisco Working Paper Series*, No. 2007–12 (2008).

Damasio, Antonio R. *Descartes' Error: Emotion, Reason, and the Human Brain*. Grosset/Putnam, 1994.

Darwin, Charles. *The Origin of Species*. Wordsworth Editions, 1998.

Davidson, R. J., and W. Irwin. "The Functional Neuroanatomy of Emotion and Affective Style." *Trends in Cognitive Sciences* 3 (1999): 11–21.

Dawkins, R. *The Selfish Gene*. Oxford University Press, 1989.

Day, E. "Why Reality TV Works." *Observer*, November 21, 2010, 18–24.

de Quervain, D. J.-F., et al. "The Neural Basis of Altruistic Punishment." *Science* 305 (2004): 1254–57.

de Waal, Frans. *The Age of Empathy: Nature's Lessons for a Kinder Society*. Harmony Books, 2009.

———. "The Evolution of Empathy." *Greater Good*, Fall/Winter 2005–6, 6–9.

———. "Putting the Altruism Back into Altruism: The Evolution of Empathy." *Annual Review of Psychology* 59 (2008): 279–300.

del Giudice, Marco, et al. "Programmed to Learn? The Ontogeny of Mirror Neurons." *Developmental Science* 12 (2009): 350–63.

Delavallade, A. "Research on Moon's Effect Could Illuminate Behavior." *Business Review* (Albany), June 23, 1997. Online.

Deschamps, J. C., and W. Doise. "Crossed Category Memberships in Intergroup Relations." in H. Tajfel, ed., *Differentiation Between Social Groups*, 141–58. Cambridge University Press, 1978.

Detillion, C. E., et al. "Social Facilitation of Wound Healing." *Psychoneuroendocrinology* 29, no. 8 (2004): 1004–11.

Di Pellegrino, G., et al. "Understanding Motor Events: A Neurophysiological Study." *Experimental Brain Research* 91 (1992): 176–80.

Ditzen, B., et al. "Effects of Different Kinds of Couple Interaction on Cortisol and Heart Rate Responses to Stress in Women." *Psychoneuroendocrinology* 32, no. 5 (2007): 565–74.

"Do Animals Think Like Autistic Savants?" *ScienceDaily*, February 20, 2008.

Donovan, S., et al. *The Physical and Psychological Effects of Meditation*, chapter 3. Institute of Noetic Sciences, 1997.

Doolan, Robert. "Helpful Animals." *Creation* 17, no. 3 (1995): 10–14.

Drew, E. J., and J. M. McGuigan. "Prevention of Crime: An Overview of Gated Communities and Neighborhood Watch." International Foundation for Protection Officers, www.ifpo.org/articlebank/gatedcommunity.html.

Drexler, Madeleine. "Where DNA Meets Daily Life." *Harvard Public Health Review*, Fall 2010, 4–9.

Dugatkin, Lee Alan. *The Altruism Equation: Seven Scientists Search for the Origins of Goodness*. Princeton University Press, 2006.

———. *Cooperation Among Animals: An Evolutionary Perspective*. Oxford University Press, 1997.

Dumé, Bell. "Entanglement Heats Up." physicsworld.com, February 23, 2006.

Dunn, B. R., et al. "Concentration and Mindfulness Meditations: Unique Forms of Consciousness?" *Applied Psychophysiology and Biofeedback* 24, no. 3 (1999): 147–65.

Eisenegger, C., et al. "Prejudice and Truth About the Effect of Testosterone on Human Bargaining Behaviour." *Nature* 463 (2010): 356–59.

Elshakry, M. "Global Darwin: Eastern Enchantment." *Nature* 461, no. 2 (2009): 1200–1201.

"Emory Brain Imaging Studies Reveal Biological Basis for Human Cooperation." *ScienceDaily*, July 18, 2002. Online.

Engardio, Joel P. "Charity Begins @ Home." *San Francisco Weekly News*, January 5, 2000. Online.

"Evolution Guides Cooperative Turn-Taking, Game Theory-Based Computer Simulations Show." *ScienceDaily*, July 9, 2009.

"Exercise and Company: Fitter with Friends." *Economist*, September 17, 2009.

Fehr, E. "The Effect of Neuropeptides on Human Trust and Altruism: A Neuroeconomic Perspective." in D. Pfaff et al., eds., *Hormones and Social Behavior*, 47–56. Springer-Verlag, 2008.

Fehr, E., and U. Fischbacher. "The Nature of Human Altruism." *Nature* 425 (2003): 785–91.

Fehr, E., and K. M. Schmidt. "A Theory of Fairness, Competition and Cooperation." *Quarterly Journal of Economics* 114 (1999): 817–68.

Fehr, E., et al. "Reciprocity as a Contract Enforcement Device: Experimental Evidence." *Econometrica* 65 (1997): 437–59.

Finch, J. "Humble MD Who Is Never Knowingly Underpaid." *Guardian*, March 7, 2008.

"First Direct Recording Made of Mirror Neurons in Human Brain." *ScienceDaily*, April 13, 2010.

Fischbacher, U., et al. "Fairness, Errors and the Power of Competition." *Journal of Economic Behavior and Organization* 72 (2009): 527–45.

Fischer, A., et al. "Recovery of Learning and Memory Is Associated with Chromatin Remodelling." *Nature* 447 (May 10, 2007): 178–82.

Fixico, Donald L. *The American Indian Mind in a Linear World: American Indian Studies and Traditional Knowledge.* Routledge, 2003.

Fogassi, L., et al. "Parietal Lobe: From Action Organization to Intention Understanding." *Science* 308 (2005): 662–67.

Forte, M., and J. X. Mitrovica. "A Resonance in the Earth's Obliquity and Precession over the Past 20 Myr Driven-by-Mantle Convection." *Nature* 390 (1997): 676–80.

Fowler, J. H., and N. A. Christakis. "Cooperative Behavior Cascades in Human Social Networks." *Proceedings of the National Academy of Sciences of the United States of America*, March 8, 2010, www.pnas.org/cgi/doi/10.1073/pnas.0913149107.

Fowler, J. H., and N. A. Christakis. "Dynamic Spread of Happiness in a Large Social Network: Longitudinal Analysis over 20 Years in the Framingham Heart Study." *British Medical Journal* 337 (2008): a2338.

Fraser, A., et al. "Association of Maternal Weight Gain in Pregnancy with Offspring Obesity and Metabolic and Vascular Traits in Childhood." *Circulation* 121 (2010): 2557-64.

Freedman, J. "White Paper: Emotional Contagion." *Sixseconds*, April 14, 2007. Online.

Friedman, H., et al. "Geomagnetic Parameters and Psychiatric Hospital Admissions." *Nature* 200 (1963): 626–28.

Fromm, E. *The Anatomy of Human Destructiveness.* Fawcett Books, 1973.

Gallese, V. "The 'Shared Manifold' Hypothesis: From Mirror Neurons to Empathy." *Journal of Consciousness Studies* 8 (2001): 33–50.

Gallese, V., et al. "Action Recognition in the Premotor Cortex." *Brain* 119, no. 2 (1996): 593–609.

Gallese, V., et al. "Intentional Attunement: Mirror Neurons and the Neural Underpinnings of Interpersonal Relations." *Journal of the American Psychoanalytic Society* 55, no. 1 (2007): 131–75.

Gallese, V., et al. "The Mirror Matching System: A Shared Manifold for Intersubjectivity." *Behavioral and Brain Sciences* 25, no. 1 (2002): 35–36.

Gallese, V., et al. "A Unifying View of the Basis of Social Cognition." *Trends in Cognitive Sciences* 8, no. 9 (2004): 396–403.

Gardner, A., and A. Grafen. "Capturing the Superorganism: A Formal Theory of Group Adaptation." *Journal of Evolutionary Biology* 22, no. 4 (2009): 659–71.

Gazzola, V., et al. "The Anthropomorphic Brain: The Mirror Neuron System Responds to Human and Robotic Actions." *NeuroImage* 35 (2007): 1674–84.

Gazzola, V., et al. "Aplasics Born Without Hands Mirror the Goal of Hand Actions with Their Feet." *Current Biology* 17 (2007): 1235–40.

Gazzola, V., et al. "Empathy and the Somatotopic Auditory Mirror System in Humans." *Current Biology* 16 (2006): 1824–29.

A Gentleman. "Smile Please." *Guardian*, October 19, 2004. Online.

Gershon, David. *Social Change 2.0: A Blueprint for Reinventing Our World.* High Point, 2009.

Gerzon, M. *Global Citizens.* Harvard Business School Press, 2006.

———. *Leading Through Conflict: How Successful Leaders Transform Differences into Opportunities.* Harvard Business School Press, 2006.

Gibson, L. "Mirrored Emotion." *University of Chicago Magazine* 98, no. 4 (2006). Online.

Gintis, H., et al. "Explaining Altruistic Behavior in Humans." *Evolution and Human Behavior* 24 (2003): 153–72.

Gislén, A. "Superior Underwater Vision in a Human Population of Sea Gypsies." *Current Biology* 13 (2003): 833–36.

Gislén, A., et al. "Visual Training Improves Underwater Vision in Children." *Vision Research* 46, no. 20 (2006): 3443–50.

Gladwell, Malcolm. *The Outliers: The Story of Success.* Penguin, 2009.

———. *The Tipping Point: How Little Things Can Make a Big Difference.* Little, Brown, 2002.

Gnevyshev, M. N. "Essential Features of the 11-Year Solar Cycle." *Solar Physics* 51 (1977): 175–82.

Goddard, G. V. "Development of Epileptic Seizures Through Brain Stimulation at Low Intensity." *Nature* 214 (1967): 1020.

Goodenough, W. "Navigation in the Western Carolines: A Traditional Science." In Laura Nadar, ed., *Naked Science: Anthropological Inquiry into Boundaries, Power, and Knowledge.* Routledge, 1996.

Goodenough, W. H., and Stephen D. Thomas. "Traditional Navigation in the Western Pacific: A Search for Pattern." *Expedition* 29, no. 3 (1987): 3–14.

Gøtzsche, P. C., et al. "Ghost Authorship in Industry-Initiated Randomised Trials." *PLoS Medicine* 4, no. 1 (2007): e19.

BIBLIOGRAPHY

Gould, Stephen Jay. *The Richness of Life.* Vintage Books, 2007.

Gramling, R., et al. "Hormone Replacement Therapy, Family History and Incident Invasive Breast Cancer Among Postmenopausal Women in the Women's Health Initiative." *Epidemiology* 20 (2009): 752–56.

Grandin, T. *Animals in Translation.* Scribner, 2005.

Greene, J., et al. "An fMRI Investigation of Emotional Engagement in Moral Judgment." *Science* 293, no. 5537 (2001): 2105–108.

Grinberg-Zylberbaum, J., et al. "The Einstein-Podolsky-Rosen Paradox in the Brain: The Transferred Potential." *Physics Essays* 7, no. 4 (1994): 422–28.

Gurfinkel, I., et al. "Assessment of the Effect of a Geomagnetic Storm on the Frequency of Appearance of Acute Cardiovascular Pathology." *Biofizika* 43, no. 4 (1998): 654–58.

Gutierrez-Garcia, J. M., and F. Tusell. "Suicides and the Lunar Cycle." *Psychological Reports* 80, no. 1 (1997): 243–50.

Haisch, B., A. Rueda, and H. E. Puthoff. "Advances in the Proposed Electromagnetic Zero-Point Field Theory of Inertia." Paper presented at AIAA 98-3143, Advances ASME/SAE/ASEE Joint Propulsion Conference and Exhibit, Cleveland, Ohio, July 13–15 1998.

Haisch, B., A. Rueda, and H. E. Puthoff. "Inertia as a Zero-Point-Field Lorentz Force." *Physical Review A* 49, no. 2 (1994): 678–94.

Haisch, Bernhard, Alfonso Rueda, and H. E. Puthoff. "Physics of the Zero-Point Field: Implications for Inertia, Gravitation and Mass." *Speculations in Science and Technology* 20 (1997): 99–114.

Halberg, F. "Chronomics of Autism and Suicide." *Biomedicine and Pharmacotherapy* 59 (2005): S100–108.

———. "Historical Encounters Between Geophysics and Biomedicine Leading to the Cornélissen-Series and Chronoastrobiology." in W. Schröder, ed., *Long- and Short-Term Variability in Sun's History and Global Change,* 272. Science Edition [2465], 2000.

Halberg, F., et al. "Cross-Spectrally Coherent about 10.5- and 21-Year Biological and Physical Cycles, Magnetic Storms and Myocardial Infarctions." *Neuroendocrinology Letters* 21 (2000): 233–58.

Halberg, F., et al. "Cycles Tipping the Scale Between Death and Survival (='Life')." *Progress of Theoretical Physics,* suppl. 173 (2008): 153–81.

Halberg, F., et al. "Subtraction and Addition of Heliomagnetics and Human Violence: Time Course of Transyears in Terrorism Mimics Their Cosmos." Presentation at the first International Workshop, Physiology of Adaptation and Quality of Life: Problems of Traditional Medicine and Innovation, People's Friendship University of Russia, Moscow, May 14–16, 2008.

Hall, Barry G. "Adaptive Evolution That Requires Multiple Spontaneous Mutations, 1. Mutations Involving an Insertion Sequence." *Genetics* 120 (1988): 887–97.

Hall, E. T. *Beyond Culture.* Anchor Books, 1989.

Halpern, David. *Social Capital.* Polity Press, 2005.

Han, S., et al. "Empathic Neural Responses to Others' Pain Are Modulated by Emotional Contexts." *Human Brain Mapping* 30 (2009): 3227–37.

Haslam, S. A., and S. D. Reicher. "The Psychology of Tyranny." *Scientific American Mind* 16, no. 3 (2005): 44–51.

Hatfield, E., et al. *Emotional Contagion*. Cambridge University Press, 1994.

Hatfield, E., et al. "Primitive Emotional Contagion: Recent Research." In J. Decety and W. Ickes, eds., *The Social Neuroscience of Empathy*. MIT Press, 2011.

Hawton, A. "The Impact of Social Isolation on the Health Status and Health-Related Quality of Life of Older People." *Quality of Life Research*, July 2010. Online. DOI: 10.1007/s11136-010-9717-2.25.

He, J. "Body Mass and Blood Pressure in a Lean Population in Southwestern China." *American Journal of Epidemiology* 139, no. 4 (1994): 380–89.

———. "Dietary Macronutrients and Blood Pressure in Southwestern China." *Journal of Hypertension* 13, no. 11 (1995): 1267–74.

———. "Effect of Migration and Related Environmental Changes on Serum Lipid Levels in Southwestern Chinese Men." *American Journal of Epidemiology* 144, no. 9 (1996): 839–48.

Heinkel-Wolfe, Peggy. "The Story Behind 'Invictus': Man Aided Team Now on Screen." *Denton Record-Chronicle* (Denton, Texas), December 24, 2009.

Henrich, J., et al. "Cooperation, Reciprocity and Punishment in Fifteen Small-Scale Societies." *American Economic Review* 91 (2001): 73–78.

Hewstone, Miles. "Mixed Communities: How 'Intergroup Contact' Can Promote Cohesion and Integration." Presentation to the Housing and Community Cohesion Conference, London, March 15, 2007.

Hochschild, A. "The Chauffeur's Dilemma." *Greater Good*, Fall/Winter 2005–6, 10–12.

Holt-Lunstad, J., et al. "Social Relationships and Mortality Risk: A Meta-analytic Review." *PLoS Medicine* 7, no. 7 (2010): e1000316.

Houston, Jean. *Jump Time: Shaping Your Future in a World of Radical Change*. Sentient Publications, 2004.

"How Chimps Deal with Death: Studies Offer Rare Glimpses." *ScienceDaily*, April 27, 2010.

Hubbard, Barbara Marx. *Conscious Evolution: Awakening the Power of Our Social Potential*. New World Library, 1998.

"Human-like Altruism Shown in Chimpanzees." *ScienceDaily*, June 25, 2007.

Hutton, W. "Fairness, Not Capitalism, Is the Issue." *Guardian*, February 26, 2010. Online.

Iacoboni, M., et al. "Grasping the Intentions of Others with One's Own Mirror Neuron System." *PLoS Biology* 3, no. 3 (2005): e79.

Institute of Noetic Sciences. *The 2007 Shift Report: Evidence of a World Transforming*. IONS Press, 2007.

———. *The 2008 Shift Report: Changing the Story of Our Future*. IONS Press, 2008.

Ispas, Alexa. "Psyched Out by Numbers: Altruism and the Dangers of Methodolatry." *EJOP (Europe's Journal of Psychology)*, August 17, 2007. Online.

Ivanoff, J. "Sea Gypsies of Myanmar." *National Geographic*, April 2005. Online.

Jabbi, M., et al. "Empathy for Positive and Negative Emotions in the Gustatory Cortex." *Chemistry and Biology* 34, no. 4 (2007): 1744–53.

Jacques, V., et al. "Experimental Realization of Wheeler's Delayed-Choice Gedankenexperiment." *Science* 315, no 5814 (2007): 966–68.

Jensen, K. "Chimpanzees Are Rational Maximizers in an Ultimatum Game." *Science* 318, no. 5847 (2007): 107–9.

Jetten, J., et al. "The Social Cure?" *Scientific American Mind*, September/October 2009.

Jirtle, R. L. "Epigenome: The Program for Human Health and Disease." *Epigenomics* 1, no. 1 (2009): 13–16.

Jirtle, Randy L., and Jennifer R. Weidman. "Imprinted and More Equal." *American Scientist,* March–April 2007, 143–49.

Johnson, David W., and Johnson, Roger T. "An Educational Psychology Success Story: Social Independence, Theory, and Cooperative Learning." *Educational Researcher* 38 (2009): 365–79.

Johnson, Sharon. "Healing Power of Community." *Oregon Mail Tribune,* March 3, 2009.

Jones, P. K., and S. L. Jones. "Lunar Association with Suicide." *Suicide and Life-Threatening Behavior* 7, no. 1 (1977): 31–39.

Kaati, G., et al. "Transgenerational Response to Nutrition, Early Life Circumstances and Longevity." *European Journal of Human Genetics* 15 (2007): 784–90.

Kanas, Nick. "Notes for the Underground." Op-ed. *New York Times,* August 29, 2010.

Kaplan, G. A., et al. "Social Connections and Mortality from All Causes and from Cardiovascular Disease: Prospective Evidence from Eastern Finland." *American Journal of Epidemiology* 128 (1988): 370–80.

Kawachi, I., et al. "Social Capital, Income Inequality, and Mortality." *American Journal of Public Health* 87, no. 9 (1997): 1491–98.

Kelly, J. R., and S. G. Barsade. "Mood and Emotions in Small Groups and Work Teams." *Organizational Behavior and Human Decision Processes* 86, no. 1 (2001): 99–130.

Keltner, Dacher. *Born to Be Good: The Science of a Meaningful Life.* Norton, 2009.

———. "The Compassionate Instinct." *Greater Good* 1 (2004): 6–9.

Keysers, C., and L. Fadiga. "The Mirror Neuron System: New Frontiers." *Social Neuroscience* 3, nos. 3–4 (2008): 193–98.

Keysers, C., and V. Gazzola. "Towards a Unifying Neural Theory of Social Cognition." In *Progress in Brain Research,* ed. S. Anders et al., 156: 379–401. Elsevier, 2006.

Keysers, C., et al. "Audiovisual Mirror Neurons and Action Recognition." *Experimental Brain Research* 153 (2003): 628–36.

Keysers, C., et al. "A Touching Sight: SII/VV Activation During the Observation and Experience of Touch." *Neuron* 42 (2004): 335–46.

Kiecolt-Glaser, J. K., et al. "Hostile Marital Interactions, Proinflammatory Cytokine Production, and Wound Healing." *Archives of General Psychiatry* 62, no. 12 (2005): 1377–84.

Kiecolt-Glaser, J. K., et al. "Marital Stress: Immunologic, Neuroendocrine, and Autonomic Correlates." *Annals of the New York Academy of Sciences* 840 (1998): 656–63.

Kiecolt-Glaser, J. K., et al. "Negative Behavior During Marital Conflict Is Associated with Immunological Down-Regulation." *Psychosomatic Medicine* 55 (1993): 395–409.

Kittenis, M., et al. "Distant Psychophysiological Interaction Effects Between Related and Unrelated Participants." *Proceedings of the Parapsychological Association Convention,* 2004, 67–76, as reported in D. Radin, *Entangled Minds,* 138–39. Paraview, 2006.

Klag, M. J. "The Contribution of Urinary Cations to the Blood Pressure Differences Associated with Migration." *American Journal of Epidemiology* 142, no. 3 (1995): 295–303.

Knight, Will. "Brain Scans Reveal Men's Pleasure in Revenge." *New Scientist,* January 18, 2006. Online.

Kollerstrom, N., and B. Steffert. "Sex Difference in Response to Stress by Lunar Month: A Pilot Study of Four Years' Crisis-Call Frequency." *BMC Psychiatry* 3 (2003): 20.

Kollewe, J., and Z. Wood. "John Lewis Staff to Share £151m in Bonuses." *Guardian,* March 11, 2010. Online.

Koski, S. E., and E. H. M. Sterchk. "Post-conflict Third-Party Affiliation in Chimpanzees: What's in It for the Third Party?" *American Journal of Primatology* 71, no. 5 (2009): 409–18. Online.

Kowinski, W. S. "Pioneers of the Heart." *North Coast Journal* (Humboldt, California), April 24, 2003. Online.

Krakovsky, M. "When Feelings Go to Work." *Wharton Alumni Magazine,* Spring 2006, 24–26.

Krivelyova, A., and C. Robotti. "Playing the Field: Geomagnetic Storms and the Stock Market." *Federal Reserve Bank of Atlanta Working Paper* 2003-5b, October 2003.

Lamm, C., et al. "The Neural Substrate of Human Empathy: Effects of Perspective-Taking and Cognitive Appraisal." *Journal of Cognitive Neuroscience* 19 (2007): 42–58.

Langergraber, K. E., et al. "The Limited Impact of Kinship on Cooperation in Wild Chimpanzees." *Proceedings of the National Academy of Sciences of the United States of America* 104 (2007): 7786–90.

Lanzetta, J. T., and B. G. Englis. "Expectations of Cooperation and Competition and Their Effects on Observers' Vicarious Emotional Responses." *Journal of Personality and Social Psychology* 56, no. 4 (1989): 543–54.

Laughlin, Robert B. *A Different Universe: Reinventing Physics from the Bottom Down.* Basic Books, 2005.

Lewis, Thomas, et al. *A General Theory of Love.* Vintage, 2000.

Lieber, A. "Human Aggression and the Lunar Synodic Cycle." *Journal of Clinical Psychiatry* 39, no. 5 (1978): 385–92.

Lieber, A., and C. R. Sherin. "Homicides and the Lunar Cycle: Toward a Theory of Lunar Influence on Human Behavior." *American Journal of Psychiatry* 129 (1972): 69–74.

Lindenberger, U., et al. "Brains Swinging in Concert: Cortical Phase Synchronization while Playing Guitar." *BMC Neuroscience* 10 (2009): 22.

Lipton, Bruce. *The Biology of Belief: Unleashing the Power of Consciousness, Matter, and Miracles.* Hay House, 2008.

Lipton, Bruce H., and Steve Bhaerman. *Spontaneous Evolution: Our Positive Future (and a Way to Get There from Here).* Hay House, 2009.

"Loneliness Can Be Contagious." *ScienceDaily,* December 2, 2009. Online.

Long, T., et al. "Relationship of Daily Geomagnetic Activity to the Occurrence of Temporal Lobe Seizures in an Epilepsy Monitoring Unit." Paper presented at American Epilepsy Society, May 1996. Abstract published in *Epilepsia,* 36(S4) (1996): 94.

Lumey, L. H. "Decreased Birthweights in Infants After Maternal In Utero Exposure to the Dutch Famine of 1944–1945." *Paediatric and Perinatal Epidemiology* 6, no. 2 (1992): 240–53.

Luscombe, B. "Do We Need $75,000 a Year to Be Happy?" *Time,* September 6, 2010.

Lutz, A., et al. "Attention Regulation and Monitoring in Meditation." *Trends in Cognitive Science* 12, no. 4 (2008): 163–69.

Lutz, A., et al. "Regulation of the Neural Circuitry of Emotion by Compassion Meditation: Effects of Meditative Expertise." *PLoS One* 3, no. 3 (2008): e1897. Online.

Mack, A., and I. Rock. "Inattentional Blindness: An Overview." *Current Directions in Psychological Science* 12, no. 5 (2003): 180–84.

Manning, R., et al. "The Kitty Genovese Murder and the Social Psychology of Helping: The Parable of the 38 Witnesses." *American Psychologist* 62 (2007): 555–62.

Markey, Sean. "Monkeys Show Sense of Fairness, Study Says." *National Geographic News*, September 17, 2003.

Marmot, Michael G., and S. Leonard Syme. "Acculturation and Coronary Heart Disease in Japanese-Americans." *American Journal of Epidemiology* 104, no. 3 (1976): 225–47.

Marsh, Jason. "The Making of an Upstander." *Greater Good* 3 (2006–7). Online.

Masson, Jeffrey, and Susan McCarthy. *When Elephants Weep: The Emotional Lives of Animals.* Vintage, 1994.

Masuda, T., and R. E. Nisbett. "Attending Holistically vs. Analytically: Comparing the Context Sensitivity of Japanese and Americans." *Journal of Personality and Social Psychology* 81 (2001): 922–34.

Mathew, V. M., et al. "Attempted Suicide and the Lunar Cycle." Pt. 1. *Psychological Reports* 68, no. 3 (1991): 927–30.

McCarthy, R., and G. Kiser. "Number of U.S. Millionaires Soared in 2009." *Huffington Post*, May 9, 2010.

McConnachie, J. "The Political Gene: How Darwin's Ideas Changed Politics. By Dennis Sewell." *Sunday Times* (London), November 8, 2009. Online.

McCraty, Rollin, et al. "The Electricity of Touch: Detection and Measurement of Cardiac Energy Exchange Between People." In Karl H. Pribram, ed., *Brain and Values: Is a Biological Science of Values Possible?*, 359–79. Lawrence Erlbaum, 1998.

McGowan, P. O., et al. "Diet and the Epigenetic (Re)programming of Phenotypic Differences in Behavior." *Brain Research* 1237 (2008): 12–24.

McGowan, P. O., et al. "Promoter-Wide Hypermethylation of the Ribosomal RNA Gene Promoter in the Suicide Brain." *PLoS One* 3, no. 5 (2008): e2085.

McMinn, D. "The Sun, the Moon, and the Number 56." *TSAA Newsletter* (Technical Securities Analysts Association of San Francisco), April 1996. Online.

McPherson, M., et al. "Birds of a Feather: Homophily in Social Networks." *Annual Review of Sociology* 27 (2001): 415–44.

McTaggart, Lynne. *The Field: The Quest for the Secret Force of the Universe.* HarperCollins, 2002.

———. *The Intention Experiment: Using Your Thoughts to Change Your Life and the World.* Free Press, 2007.

Meaney, M. J., and M. Szyf. "Environmental Programming of Stress Responses Through DNA Methylation: Life at the Interface Between a Dynamic Environment and a Fixed Genome." *Dialogues in Clinical Neuroscience* 7, no. 2 (2005): 103–23.

"Memory Restored in Mice Through Enriched Environment: New Hope for Alzheimer's." *ScienceDaily*, April 30, 2007.

BIBLIOGRAPHY

Meredith, C. "Tit for Tat." *The Slab* (Australian Broadcasting Corporation), 1998. Online.

Michon, A., and M. Persinger. "Experimental Simulation of the Effects of Increased Geomagnetic Activity upon Nocturnal Seizures in Epileptic Rats." *Neuroscience Letters* 224 (1997): 53–56.

Mikulecky, M. "Lunisolar Tidal Waves, Geomagnetic Activity and Epilepsy in the Light of Multivariate Coherence." *Brazilian Journal of Medical and Biological Research* 29 (1996): 1069–72.

Mill, J., et al. "Epigenomic Profiling Reveals DNA-Methylation Changes Associated with Major Psychosis." *American Journal of Human Genetics* 82, no. 3 (2008): 696–711.

Mills, Kelly. "Bringing the 'Social' to Epidemiology: S. Leonard Syme, Ph.D." *Public Health Magazine*, University of California at Berkeley, Spring 2006: 22–23.

Misra, N. "Stone Age Cultures Survive Tsunami Waves." Associated Press, January 4, 2005. Online.

Mizun, Y. G., and P. G. Mizun. *Space and Health.* Znanie, 1984.

Moll, J., et al. "Human Fronto-Mesolimbic Networks Guide Decisions about Charitable Donation." *Proceedings of the National Academy of Sciences of the United States of America* 103 (2006): 15623–28.

Monroe, Kristen Renwick. *The Heart of Altruism: Perceptions of a Common Humanity.* Princeton University Press, 1996.

Mott, M. "Did Animals Sense Tsunami Was Coming?" *National Geographic News*, January 4, 2005. Online.

Moyer, M. W. "You're Happy, I'm Happy." *Scientific American Mind*, September/October 2010, 13.

Müller-Fahrenholz, Geiko. *The Art of Forgiveness: Theological Reflections on Healing and Reconciliation.* WCC Publications, 1997.

Murphy, M., et al. *The Physiological and Psychological Effects of Meditation: A Review of Contemporary Research, with a Comprehensive Bibliography, 1931–1996.* Institute of Noetic Sciences, 1997.

Myers Lowe, R. "Combating Childhood Obesity May Start in the Womb." Reuters, May 14, 2010. Online.

Naish, John, *Enough: Breaking Free from the World of Excess.* Hodder and Stoughton, 2009.

Nakamura, K., and M. Hiramatsu. "Ultra-Weak Photon Emission from Human Hand: Influence of Temperature and Oxygen Concentration on Emission." *Journal of Photochemistry and Photobiology B: Biology* 80, no. 2 (2005): 156–60.

Neal, R. D., and M. Colledge. "The Effect of the Full Moon on General Practice Consultation Rates." *Family Practice* 17, no. 6 (2000): 472–74.

Neumann, R., and F. Strack. "'Mood Contagion': The Automatic Transfer of Mood Between Persons." *Journal of Personality and Social Psychology* 79 (2000): 211–23.

Neumann, R., et al. "The Influence of Mood on the Intensity of Emotional Responses: Disentangling Feeling and Knowing." *Cognition and Emotion* 15, no. 6 (2001): 725–47.

"The New 'Epigenetics': Poor Nutrition in the Womb Causes Permanent Genetic Changes in Offspring." *ScienceDaily*, April 14, 2009.

Newberg, A. "Cerebral Blood Flow During Meditative Prayer: Preliminary Findings and Methodological Issues." *Perceptual and Motor Skills* 97 (2003): 625–30.

Newberg, A. B., and J. Iversen. "The Neural Basis of the Complex Mental Task of Meditation: Neurotransmitter and Neurochemical Considerations." *Medical Hypotheses* 61, no. 2 (2003): 282–91.

Newman, M. E. J. "The Structure and Function of Complex Networks." *SIAM Review* (Society for Industrial and Applied Mathematics) 45 (2003): 167–256.

Nicolson, S. "Child Vampire Hunters Sparked Comic Crackdown." *BBC News,* March 22, 2010. Online.

Nisbett, Richard E. *The Geography of Thought: How Asians and Westerners Think Differently—and Why.* Nicholas Brealey, 2009.

Nisbett, R., and T. Masuda. "Culture and Point of View." *Proceedings of the National Academy of Sciences of the United States of America* 100, no. 19 (2003): 11163–70.

Nitsche, M. "Are the Stabilizing and Destabilizing Influences of the Planetary Gravitational Field on the Structural Formation of Biological Patterns Real?" Lecture at the 10th Conference on Synergetics and Complexity Research, Self-Organization in Psychology, Psychiatry and Social Sciences, Bavaria, Germany, June 6–8, 2002.

Nitschke, J. B., et al. "Orbitofrontal Cortex Tracks Positive Mood in Mothers Viewing Pictures of Their Newborn Infants." *NeuroImage* 21, no. 2 (2004): 583–92.

Oatley, K. "A Feeling for Fiction." *Greater Good,* Fall/Winter 2005–6, 12–15.

O'Connell, S. M. "Empathy in Chimpanzees: Evidence for Theory of Mind?" *Primates* 36, no. 3 (1995): 397–410.

O'Connor, R. P., and M. A. Persinger. "Geophysical Variables and Behavior LXXXII: A Strong Association Between Sudden Infant Death Syndrome (SIDS) and Increments of Global Geomagnetic Activity. Possible Support for the Melatonin Hypothesis." *Perceptual and Motor Skills* 84 (1997): 395–402.

O'Dea, James. *Creative Stress: A Path for Evolving Souls Living Through Personal and Planetary Upheaval.* Self-published, 2010.

Oderda, G. M., and W. Klein-Schwartz. "Lunar Cycle and Poison Center Calls." *Journal of Toxicology and Clinical Toxicology* 20, no. 5 (1983): 487–95.

Oliner, S. *Do unto Others: Extraordinary Acts of Ordinary People.* Westview Press, 2003.

Oraevskii, N., et al. "Medico-Biological Effect of Natural Electromagnetic Variations." *Biofizika* 43, no. 5 (1998): 844–48.

Oraevskii, V. N., et al. "An Influence of Geomagnetic Activity on the Functional Status of the Body." *Biofizika* 43, no. 5 (1998): 819–26.

Ormel, Johan, et al. "Vulnerability Before, During, and After a Major Depressive Episode: A 3-Wave Population-Based Study." *Archives of General Psychiatry* 61 (2004): 990–96.

Ou, J. N., et al. "Histone Deacetylase Inhibitor Trichostatin A Induces Global and Gene-Specific DNA Demethylation in Human Cancer Cell Lines." *Biochemical Pharmacology* 73, no. 9 (2007): 1297–307.

Page, L. B., et al. "Antecedents of Cardiovascular Disease in Six Solomon Island Societies." *Circulation* 49 (1974): 1132–46.

Panksepp, J. "The Anatomy of Emotions." In R. Plutchik, ed., *Emotion: Theory, Research and Experience,* vol. 3: *Biological Foundations of Emotions,* 91–124. Academic Press, 1986.

———. "The Neurobiology of Emotions: Of Animal Brains and Human Feelings." in T. Manstead and H. Wagner, eds., *Handbook of Psychophysiology,* 5–26. Wiley, 1989.

BIBLIOGRAPHY

Park, Alice. "Study: Money Isn't Everything—But Status Is!" *Time*, March 23, 2010. Online.

Pembrey, M. E., et al. "Sex-Specific, Male-Line Transgenerational Responses in Humans." *European Journal of Human Genetics* 14 (2006): 159–66.

Persinger, M. "Sudden Unexpected Death in Epileptics Following Sudden, Intense, Increases in Geomagnetic Activity: Prevalence of Effect and Potential Mechanisms." *International Journal of Biometeorology* 38 (1995): 180–87.

Petro, M., et al. "An Influence of Changes of Magnetic Field of the Earth on the Functional State of Humans in the Conditions of Space Mission." Paper presented at the International Symposium Computer Electro-Cardiograph on Boundary of Centuries, Moscow, April 27–30, 1999.

Pettigrew, T. F. "Intergroup Contact Theory." *Annual Review of Psychology* 49 (1998): 65–85.

Pettigrew, T. F., and L R. Tropp. "A Meta-analytic Test of Intergroup Contact Theory." *Journal of Personality and Social Psychology* 90, no. 5 (2006): 751–83.

Phillips, N. J., et al. "Absence of Pupil Response to Blur-Driven Accommodation." *Vision Research* 32, no. 9 (1992): 1775–79.

Pichon, S., et al. "Emotional Modulation of Visual and Motor Areas by Dynamic Body Expressions of Anger." *Social Neuroscience* 3, nos. 3–4 (2008): 199–212.

Piliavin, M. "Good Samaritanism: An Underground Phenomenon?" *Journal of Personality and Social Psychology* 13, no. 4 (1969): 289–99.

Pitzke, M. "The World As We Know It Is Going Down." *Spiegel*, September 18, 2008. Online.

Popp, F.-A., et al. "Mechanism of Interaction Between Electromagnetic Fields and Living Organisms." *Science in China* (Series C), 43, no. 5 (2000): 507–18.

Post, R. M. "Transduction of Psychosocial Stress into the Neurobiology of Recurrent Affective Disorder." *American Journal of Psychiatry* 149 (1992): 999–1010.

Post, S. G. "Altruism, Happiness and Health: It's Good to Be Good." *International Journal of Behavioral Medicine* 12, no. 2 (2005): 66–77.

Potter, M., and J. Davey. "M&S Profit Rise Lags Rivals." Reuters, May 25, 2010. Online.

Pray, Leslie A. "Epigenetics: Genome, Meet Your Environment." *Scientist* 18, no. 13 (2004): 14–20.

Preston, S. D., and F. B. M. de Waal. "Empathy: Its Ultimate and Proximate Bases." *Behavioral and Brain Sciences* 25 (2002): 1–72.

Preston, S. D., et al. "The Neural Substrates of Cognitive Empathy." *Social Neuroscience* 2, nos. 3–4 (2007): 254–75.

"Public Conversations Project, 'Talking with the Enemy.'" *Boston Globe*, January 28, 2001.

Pugh, S. D. "Service with a Smile: Emotional Contagion in the Service Encounter." *Academy of Management Journal* 44, no. 5 (2001): 1018–27.

Pusey, J. "Global Darwin: Revolutionary Road." *Nature* 462, no. 12 (2009): 162–63.

Puthoff, H. E. "Ground State of Hydrogen as a Zero-Point-Fluctuation-Determined State." *Physical Review D* 35 (1987): 3266.

Putnam, Robert D. *Bowling Alone: The Collapse and Revival of American Community*. Simon & Schuster, 2000.

———. "E Pluribus Unum: Diversity and Community in the Twenty-First Century:

The 2006 Johan Skytte Prize Lecture." *Scandinavian Political Studies* 30, no. 2 (2007): 137–74.

———. "Social Capital: Measurement and Consequences." *ISUMA: Canadian Journal of Policy Research* 2, no. 1 (2001): 41–51.

Putnam, R. D., and David E. Campbell. *American Grace: How Religion Divides and Unites Us.* Simon & Schuster, 2010.

Putnam, Robert D., and Lewis M. Feldstein. *Better Together: Restoring the American Community.* Simon & Schuster, 2003.

Putnam, Robert, et al. "The Social Capital Community Benchmark Survey." www. hks.harvard.edu/saguaro/communitysurvey/results_pr.html.

Radin, D. I. "Event-Related EEG Correlations Between Isolated Human Subjects." *Journal of Alternative and Complementary Medicine* 10 (2004): 315–24.

Radin, D. I., and M. J. Schlitz. "Gut Feelings, Intuition and Emotions: An Exploratory Study." *Journal of Alternative and Complementary Medicine* 11, no. 5 (2005): 85–91.

Ramkissoon, Anton. "Karma Kitchen in Northwest D.C. Serves Trust and Generosity." *Washington Post,* May 6, 2010. Online.

Rand, Ayn. *The Fountainhead.* Signet, 1996.

Raps, Avi, et al. "Geophysical Variables and Behavior: LXIX. Solar Activity and Admission of Psychiatric Inpatients." *Perceptual and Motor Skills* 74 (1992): 449.

Rasenberger, J. "Nightmare on Austin Street." *American Heritage,* October 5, 2006. Online.

Reed, D., et al. "Social Networks and Coronary Heart Disease Among Japanese Men in Hawaii." *American Journal of Epidemiology* 117 (1983): 384–96.

Reiman, J. "The Impact of Meditative Attentional Training on Measures of Select Attentional Parameters and on Measures of Client-Perceived Counsellor Empathy." *Dissertation Abstracts International* 46, no. 6A (1985): 1569.

"Relationships Improve Your Odds of Survival by 50 Percent, Research Finds." *ScienceDaily,* July 27, 2010.

Rettner, R. "Brain's 'Fairness' Spot Found." *LiveScience,* February 24, 2010. Online.

Rice, G. E., and P. Gainer. "'Altruism' in the Albino Rat." *Journal of Comparative and Physiological Psychology* 55 (1962): 123–25.

Rifkin, Jeremy. *The Empathic Civilization: The Race to Global Consciousness in a World in Crisis.* Polity, 2009.

Riley, Naomi Schaefer. "Getting to Know You." *Wall Street Journal,* May 15, 2009.

Rilling, J., et al. "A Neural Basis for Social Cooperation." *Neuron* 35, no. 2 (2002): 395–405.

Rimé, Bernard, and E. Zech. "The Social Sharing of Emotions; Interpersonal and Collective Dimensions." *Boletin de Psicologia* 70 (2001): 97–108.

Rimé, Bernard, et al. "Beyond the Emotional Event: Six Studies on the Social Sharing of Emotion." *Cognition and Emotion* 5, nos. 5–6 (1991): 435–65.

Rizzolatti, G., and L. Craighero. "Mirror Neuron: A Neurological Approach to Empathy." In Jean-Pierre Changeux et al., eds., *Neurobiology of Human Values,* 108–23. Springer-Verlag, 2005.

Rizzolatti, G., and L. Craighero. "The Mirror Neuron System." *Annual Review of Neuroscience* 27 (2004): 169–92.

BIBLIOGRAPHY

Rizzolatti, Giacomo, and Corrado Sinigaglia. *Mirrors in the Brain: How Our Minds Share Actions and Emotions.* Oxford University Press, 2008.

Rosenblum, B., and F. Kuttner. "The Observer in the Quantum Experiment." *Foundations of Physics* 32, no. 8 (2002): 1273–93.

Rossouw, Jacques E., et al. "Risks and Benefits of Estrogen Plus Progestin in Healthy Postmenopausal Women: Principal Results from the Women's Health Initiative Randomized Controlled Trial." *Journal of the American Medical Association* 288 (2002): 321–33.

Ruby, P., and J. Decety. "How Would You Feel versus How Do You Think She Would Feel?" *Journal of Cognitive Neuroscience* 16, no. 6 (2004): 988–99.

Russell, C. T., et al. "The Permanent and Induced Magnetic Dipole Moment of the Moon." *Proceedings of the Fifth Lunar Conference* 3 (1974): 2747–60.

Sample, I. "Working in a Team Increases Human Pain Threshold." *Guardian*, September 16, 2009.

Sanders, J. V. "I Am the Son of Sam." *Fortean Times* (London), August 2002. Online.

Saul, John Ralston. *The Collapse of Globalism.* Atlantic Books, 2005.

"Sawubona." Video. Global Oneness Project. September 27, 2006.

Scanlan, J. M., et al. "CD4 and CD8 Counts Are Associated with Interactions of Gender and Psychosocial Stress." *Psychosomatic Medicine* 60, no. 5 (1998): 644–53.

Schedlowski, M., et al. "Psychophysiological, Neuroendocrine and Cellular Immune Reactions under Psychological Stress." *Neuropsychobiology* 28 (1993): 87–90.

Scherwitz, L., et al. "Self-Involvement and Coronary Heart Disease Incidence in the Multiple Risk Factor Intervention Trial." *Psychosomatic Medicine* 48 (1986): 187–99.

Scheving, L. E., and F. Halberg. *Chronobiology: Principles and Applications to Shifts in Schedules.* Kluwer Academic, 1981.

Schulte-Rüther, M., et al. "Mirror Neuron and Theory of Mind Mechanisms Involved in Face-to-Face Interactions: A Functional Magnetic Resonance Imaging Approach to Empathy." *Journal of Cognitive Neuroscience* 19, no. 8 (2007): 1354–72.

"The Sea Gypsies." *60 Minutes.* March 20, 2005.

Search for Common Ground. "Cooperative Problem-Solving: A Guide for Turning Conflicts into Agreements." 2003. Online.

Sephton, S. E., et al. "Diurnal Cortisol Rhythm as a Predictor of Breast Cancer Survival." *Journal of the National Cancer Institute* 92, no. 2 (2000): 994–1000.

Sewell, Dennis. "Charles Darwin and the Children of the Evolution." *London Times*, November 8, 2009.

Shankleman, M. "Income Envy 'Makes You Ill.'" *BBC News*, April 20, 2009.

Shapiro, J. A. "Bacteria Are Small but Not Stupid: Cognition, Natural Genetic Engineering and Socio-Bacteriology." *Studies in History and Philosophy of Science Part C: Studies in History and Philosophy of Biological and Biomedical Sciences* 38, no. 4 (2007): 807–19.

———. "Mobile DNA and Evolution in the 21st Century." *Mobile DNA* 1 (2010): 4.

———. "Revisiting the Central Dogma in the 21st Century." *Annals of the New York Academy of Sciences* 1178 (2009): 6–28.

———. "A 21st Century View of Evolution." *Journal of Biological Physics* 28 (2002): 1–20.

Shenk, Joshua Wolf. "What Makes Us Happy?" *Atlantic*, June 2009.

Shepherd, S. V., et al. "Latency of Social-Cued Attention Signals in Macaque Area LIP." Paper presented at Neuroscience 2007, the 37th annual meeting of the Society for Neuroscience, San Diego, 2007.

Sherif, M., et al. *The Robbers Cave Experiment: Intergroup Conflict and Cooperation, 1954.* University of Oklahoma Institute of Group Relations, 1961.

Sigmund, K., and C. Hauert. "Altruism," *Current Biology* 12, no. 8 (2002): R270–2.

Simons, D. J., and C. F. Chabris. "Gorillas in Our Midst: Sustained Inattentional Blindness for Dynamic Events." *Perception* 28 (1999): 1059–74.

Simons, D. J., and D. T. Levin. "Failure to Detect Changes to People During a Real-World Interaction." *Psychonomic Bulletin and Review* 5 (1998): 644–49.

Singer, T., and E. Fehr. "The Neuroeconomics of Mind Reading and Empathy." *American Economic Review* 95, no. 2 (2005): 340–45.

Singer, T., et al. "Empathic Neural Responses Are Modulated by the Perceived Fairness of Others." *Nature* 439 (2006): 466–69.

Singer, T., et al. "Empathy for Pain Involves the Affective but Not Sensory Components of Pain." *Science* 303, no. 5661 (2004): 1157–62.

Sitar, J. "The Causality of Lunar Changes on Cardiovascular Mortality." *Casopis Lekaru Ceskych* 129 (1990): 1425–30.

Slagter, H. A., et al. "Mental Training Affects Distribution of Limited Brain Resources." *PLoS Biology* 5, no. 6 (2007): e138.

Smolin, Lee. *The Trouble with Physics.* Penguin, 2006.

Snyder, A. W., et al. "Savant-Like Skills Exposed in Normal People by Suppressing the Left Fronto-Temporal Lobe." *Journal of Integrative Neuroscience* 2, no. 2 (2003): 149–58.

"Social Isolation Worsens Cancer, Mouse Study Suggests." *ScienceDaily*, September 29, 2009.

Spiegel, D., and S. E. Sephton. "Psychoneuroimmune and Endocrine Pathways in Cancer: Effects of Stress and Support." *Seminars in Clinical Neuropsychiatry* 6, no. 4 (2001): 252–65.

Stafford, N. "Laughter: It's Catching." *Nature News*, December 12, 2006. Online.

Standish, L., et al. "Electroencephalographic Evidence of Correlated Event-Related Signals Between the Brains of Spatially and Sensory Isolated Human Subjects." *Journal of Alternative and Complementary Medicine* 10, no. 2 (2004): 307–14.

Starbuck, S., et al. "Is Motivation Influenced by Geomagnetic Activity?" *Biomedicine and Pharmacotherapy* 56 (2002): 289s–97s.

Staub, E., and J. Vollhardt. "Altruism Born of Suffering: The Roots of Caring and Helping After Victimization and Other Trauma." *American Journal of Orthopsychiatry* 78, no. 3 (2008): 267–80.

Stenhouse, D. "Tales of the Gorbals Vampire." *Sunday Times* (London), March 28, 2010.

Steptoe, A., and Ana V. Diez Roux. "Happiness, Health, and Social Networks." Editorial. *British Medical Journal* 339 (2009): a2781.

Sterelny, Kim. *Dawkins vs. Gould: Survival of the Fittest.* Icon Books, 2007.

Stern, S. "How to Respond when the Truth Hurts." *Financial Times*, May 25, 2010. Online.

Stoilova, I., and T. Zdravev. "Influence of the Geomagnetic Activity on the Human Functional Systems." *Journal of the Balkan Geophysical Society* 3, no. 4 (2000): 73–76.

BIBLIOGRAPHY

Stoupel, E. "Relationship Between Suicide and Myocardial Infarction with Regard to Changing Physical Environmental Conditions." *International Journal of Biometeorology* 38, no. 4 (1994): 199–203.

Stoupel, E., et al. "Clinical Cosmobiology: The Lithuanian Study, 1990–1992." *International Journal of Biometerology* 38 (1995): 204–8.

———. "Suicide-Homicide Temporal Interrelationship, Links with Other Fatalities and Environmental Physical Activity." *Crisis* 26 (2005): 85–89.

"The Subway Samaritan." *Time*, January 19, 1970.

Sutton, R. I. "Boss Luis Urzúa and the Trapped Miners in Chile: A Classic Case of Leadership, Performance, and Humanity." *Psychology Today*, September 6, 2010.

Swan, G. E. and D. Carmelli. "Curiosity and Mortality in Aging Adults: A 5-Year Follow-Up of the Western Collaborative Group Study." *Psychology and Aging* 11 (1996): 449–53.

Syme, L. "Historical Perspective: The Social Determinants of Disease—Some Roots of the Movement." *Epidemiological Perspectives and Innovations*, 2 (2005): 2.

Syme, S. L., et al., "Cultural Mobility and the Occurrence of Coronary Heart Disease." *Journal of Health and Human Behavior* 6 (1965): 178–89.

Syme, S. L., et al., "Epidemiologic Studies of Coronary Heart Disease and Stroke in Japanese Men Living in Japan, Hawaii and California: Introduction." *American Journal of Epidemiology* 102, no. 6 (1975): 477–80.

Syme, S. L., et al., "Some Social and Cultural Factors Associated with the Occurrence of Coronary Heart Disease." *Journal of Chronic Diseases* 17 (1964): 277–89.

Szyf, M. "The Role of DNA Hypermethylation in Cancer and Cancer Therapy." *Current Oncology* 15, no. 2 (2008): 72–75.

Szyf, M., et al. "DNA Demethylation and Cancer: Therapeutic Implications." *Cancer Letters* 211, no. 2 (2004): 133–43.

Szyf, M., et al. "DNA Methylation and Breast Cancer." *Biochemical Pharmacology* 68, no. 6 (2004): 1187–97.

Szymanski, Richard. "Can Changing Neighborhood Traffic Circulation Patterns Reduce Crime and Improve Personal Safety? A Quantitative Analysis of One Neighborhood's Efforts." Unpublished master's thesis, Florida Atlantic University, Department of City and Regional Planning, 1994.

"Tailholt Residents Build Community Center," *Cherokee News Path*, March 10, 2005. Online.

Tajfel, H. "Experiments in Intergroup Discrimination." *Scientific American* 223 (1970): 96–102.

"Teenagers Hit by Soap Opera Virus." *Reuters.* May 19, 2006.

Templer D., and D. Veleber. "The Moon and Madness: A Comprehensive Perspective." *Journal of Clinical Psychology* 36, no 4 (1980): 865–68.

Thioux, M., et al. "Action Understanding: How, What and Why." *Current Biology* 18, no. 10 (2008): 431–34.

Thomas, C. B., and D. C. Ross. "Precursors of Hypertension and Coronary Disease Among Healthy Medical Students: Discriminant Function Analysis V. Family Attitudes." *Johns Hopkins Medical Journal* 123 (1968): 283–96.

Thompson, J. presenter. "On Forgiveness and Social Healing." Panel Discussion on the Role of Forgiveness in Social Healing, Harvard Divinity School, October 31, 2005. Online at www.humiliationstudies.org/documents/ThompsonForgiveness.pdf.

Tierney, Ruth. "The Search for Happiness." *Sunday Times* (London), December 20, 2009. Online.

Todes, D. "Global Darwin: Contempt for Competition." *Nature* 462, no. 5 (2009): 36–37.

Tremmel, P. V. "Culture of 'We' Buffers Genetic Tendency to Depression." *Northwestern University News*, October 27, 2009.

Tricomi, E., et al. "Neural Evidence for Inequality-Averse Social Preferences." *Nature* 463 (2010): 1089–91.

Tucker, Chris. "New Maps of Hope: Breaking Out of the Trap of Race." *D Magazine* (Dallas), October 1991.

"Tuning in to Genius." *BBC News*, October 7, 1999.

University of Leicester. "Research Shows That 'Invisible Hand' Guides Evolution of Cooperative Turn-Taking." Press release. *Science News*, July 8, 2009.

Uvnas-Moberg, K., and M. Petersson. "Oxytocin, A Mediator of Anti-stress." *Zeitschrift für Psychosomatische Medizin und Psychotherapie* 51, no. 1 (2005): 57–80.

Van de Gaag, C., et al. "Facial Expressions: What the Mirror Neuron System Can and Cannot Tell Us." *Social Neuroscience* 2, nos. 3–4 (2007): 179–222.

Van Stambrouck, Paul. "Getting Radical in California—1990s Style." charityfocus. org.

Vanderwal, T. "Everyone Around You Acting Strange? Maybe It's a Full Moon." *Olympian* (Olympia, Washington), September 15, 2003.

Vedral, V. "A Better Than Perfect Match." *Nature* 439 (2006): 397.

———. "Quantifying Entanglement in Macroscopic Systems." *Nature* 453 (2009): 1004–7.

———. "Where Does Life Come From?" *Straits Times*, January 31, 2009, 8.

Vormbrock, J. K., and J. M. Grossberg. "Cardiovascular Effects of Human–Pet Dog Interactions." *Journal of Behavioral Medicine* 11 (1988): 509–17.

Wade, Dorothy. "So What Do You Have to Do to Find Happiness?" *Sunday Times* (London), October 2, 2005. Online.

Waller, J. "Dancing Death." *BBC News*, September 12, 2008. Online.

Warneken, F., and M. Tomasello. "Altruistic Helping in Human Infants and Young Chimpanzees." *Science* 311 (2006): 1301–3.

Warneken, F., et al. "Spontaneous Altruism by Chimpanzees and Young Children." *PLoS Biology* 5, no. 7 (2007): 1414–20.

Warren, J. E., et al. "Positive Emotions Preferentially Engage an Auditory-Motor Mirror System." *Journal of Neuroscience* 26, no. 50 (2006): 13067–75.

Waterland, R. A., and R. L. Jirtle. "Transposable Elements: Targets for Early Nutritional Effects on Epigenetic Gene Regulation." *Molecular Cell Biology* 23 (2003): 5293–300.

Waterland, R. A., et al. "Methyl Donor Supplementation Prevents Transgenerational Amplification of Obesity." *International Journal of Obesity* 32, no. 9 (2008): 1373–79.

Wechkin, J. H., et al. "Altruistic Behavior in Rhesus Monkeys." *American Journal of Psychiatry* 121 (1964): 584–85.

Weaver, I. C., et al. "Epigenetic Programming by Maternal Behavior." *Nature Neuroscience* 7, no. 8 (2004): 847–54.

BIBLIOGRAPHY

Weaver, I. C., et al. "Maternal Care Effects on the Hippocampal Transcriptome and Anxiety-Mediated Behaviors in the Offspring That Are Reversible in Adulthood." *Proceedings of the National Academy of Sciences of the United States of America* 103, no. 9 (2006): 3480–85.

Weaver, I. C., et al. "Reversal of Maternal Programming of Stress Responses in Adult Offspring Through Methyl Supplementation: Altering Epigenetic Marking Later in Life." *Journal of Neuroscience* 25, no. 47 (2005): 11045–54.

Webber, J., and J. P. Rathbone. "Man in the News: Luis Urzúa." *Financial Times* (London), October 15, 2010.

Weinberg, Steven. "What Is an Elementary Particle?" *Beam Line: A Periodical of Particle Physics* 27 (Spring 1997): 17–21.

Wessely, S. "Protean Nature of Mass Sociogenic Illness." *British Journal of Psychiatry* 180 (2002): 300–306.

"What Is a Scab?" *BBC News*, March 4, 2004.

Wicker, B., et al. "Both of Us Disgusted in My Insula: The Common Neural Basis of Seeing and Feeling Disgust." *Neuron* 40 (2003): 655–64.

Wilkinson, R., and Kate Pickett. *The Spirit Level: Why More Equal Societies Almost Always Do Better.* Allen Lane/Penguin, 2009.

Williams, J. B., et al. "A Model of Gene-Environment Interactions Reveals Altered Mammary Gland Gene Expression and Increased Tumor Growth Following Social Isolation." *Cancer Prevention Research* 2 (2009): 850.

Williams, P., and M. West. "EEG Responses to Photic Stimulation in Persons Experienced at Meditation." *Electroencephalography and Clinical Neurophysiology* 39, no. 5 (1975): 519–22.

Wright, S. "Fat Cats in Terror After Anti-Capitalists Attack Fred the Shred's Home." *London Daily Mail Online*, March 26, 2009.

Wolf, S., and J. G. Bruhn. *The Power of Clan.* Transaction, 1993.

Wolf, S., et al. "Roseto Revisited: Further Data on the Incidence of Myocardial Infarction in Roseto and Neighboring Pennsylvania Communities." *Transactions of the American Clinical and Climatological Association* 85 (1973): 100–108.

Wolf, S., et al. "Roseto, Pennsylvania 25 Years Later—Highlights of a Medical and Sociological Survey." *Transactions of the American Clinical and Climatological Association* 100 (1989): 57–67.

Wolff G., et al. "Maternal Epigenetics and Methyl Supplements Affect Agouti Gene Expression in Avy/a Mice." *FASEB Journal* 12 (August 1998): 949–57.

Woodward, W. "New Surprises in Very Old Places: Civil War Nurse Leaders and Longevity." *Nursing Forum* 26, no. 1 (1991): 9–16.

Yamamoto, M., et al. "An Experiment on Remote Action against Man in Sense Shielding Condition." *Journal of the International Society of Life Information Sciences* 14, no. 1 (1996): 97–99.

Zimbardo, P. G. *The Lucifer Effect: Understanding How Good People Turn Evil.* Random House, 2007.

INDEX

INDEX

INDEX

INDEX

INDEX

INDEX

INDEX

INDEX

INDEX

INDEX

INDEX

ABOUT THE AUTHOR

Lynne McTaggart is an award-winning journalist and author of the international best-selling books *The Field* (www.lynnemctaggart.com) and *The Intention Experiment* (www.theintentionexperiment.com), featured in Dan Brown's *The Lost Symbol*. As cofounder and editorial director of the magazine *What Doctors Don't Tell You* (www.wddty.com), she publishes health newsletters that are among the most widely praised health publications in the world. She is also the editor of *Living the Field*, a course that helps to bring the science of The Field into everyday life. Her company also holds highly popular conferences and workshops on health and spirituality.

She has become a well-respected international authority on the science of spirituality.

McTaggart is also the author of *The Baby Brokers: The Marketing of White Babies in America* (Dial Press) and *Kathleen Kennedy: Her Life and Times* (Dial Press, Weidenfeld & Nicolson). *The Intention Experiment, The Field,* and What Doctors Don't Tell You have been translated into twenty-eight languages.

She and her husband, WDDTY cofounder Bryan Hubbard, live and work in London with their two daughters.

McTaggart is a featured expert in the powerful and inspirational film, *I Am* (www.iamthedoc.com), which premiered in 2011.

MAKE A DEEPER CONNECTION

The Bond and the Bond practices are only the start of a continuing journey that you can take with Lynne McTaggart (www.lynnemctaggart .com).

To find out more about *The Bond* and Bond intensive training programs for individuals or for business, see www.thebond.net.

To join Lynne's ongoing Intention Experiment and to receive weekly updates about the experiments, written by Lynne, please go to: www .theintentionexperiment.com.

Signing up will also provide you with information about Intention Experiment membership and ongoing teleseminars and our VIP membership program, enabling you to live a life of holism in every regard.

For some of the world's finest information about healthcare, modern medicine and better alternatives, to join our medical 'clubs' and teleseminars, to receive weekly updates about your health in Lynne's health blogs, and to find out how to get Lynne's magazine *What Doctors Don't Tell You*, visit www.wddty.com.

www.thebond.net